Student Support Materials for
Edexcel A Level Maths
Core 1

Authors: John Berry and Sue Langham

William Collins' dream of knowledge for all began with the publication of his first book in 1819. A self-educated mill worker, he not only enriched millions of lives, but also founded a flourishing publishing house. Today, staying true to this spirit, Collins books are packed with inspiration, innovation and practical expertise. They place you at the centre of a world of possibility and give you exactly what you need to explore it.

Collins. Freedom to teach.

Published by Collins
An imprint of HarperCollins*Publishers*
77–85 Fulham Palace Road
Hammersmith
London
W6 8JB

© HarperCollins*Publishers* Limited 2012

10 9 8 7 6 5 4 3 2 1

ISBN-13 978-0-00-747601-5

British Library Cataloguing in Publication Data. A Catalogue record for this publication is available from the British Library.

Commissioned by Lindsey Charles and Emma Braithwaite
Project managed by Lindsey Charles
Edited and proofread by Susan Gardner
Reviewed by Stewart Townend
Design and typesetting by Jouve India Private Limited
Illustrations by Ann Paganuzzi
Index compiled by Michael Forder
Cover design by Angela English
Production by Simon Moore

Printed and bound in Spain by Graficas Estella

Browse the complete Collins catalogue at:
www.collinseducation.com

This material has been endorsed by Edexcel and offers high quality support for the delivery of Edexcel qualifications.

Edexcel endorsement does not mean that this material is essential to achieve any Edexcel qualification, nor does it mean that this is the only suitable material available to support any Edexcel qualification. No endorsed material will be used verbatim in setting any Edexcel examination and any resource lists produced by Edexcel shall include this and other appropriate texts. While this material has been through an Edexcel quality assurance process, all responsibility for the content remains with the publisher.

Copies of official specifications for all Edexcel qualifications may be found on the Edexcel website - www.edexcel.com

Acknowledgements

The publishers wish to thank the following for permission to reproduce photographs. Every effort has been made to trace copyright holders and to obtain their permission for the use of copyright material. The publishers will gladly receive any information enabling them to rectify any error or omission at the first opportunity.

Cover image: Abstract Images Mirrored in Glass Wall © Lixiaofeng | Dreamstime.com

MIX
Paper from
responsible sources
FSC C007454

FSC™ is a non-profit international organisation established to promote the responsible management of the world's forests. Products carrying the FSC label are independently certified to assure consumers that they come from forests that are managed to meet the social, economic and ecological needs of present and future generations, and other controlled sources.

Find out more about HarperCollins and the environment at
www.harpercollins.co.uk/green

Welcome to Collins Student Support Materials for Edexcel A level Mathematics. This page introduces you to the key features of the book which will help you to succeed in your examinations and to enjoy your maths course.

The chapters are organised by the main sections within the specification for easy reference. Each one gives a succinct explanation of the key ideas you need to know.

Examples and answers

After ideas have been explained the worked examples in the green boxes demonstrate how to use them to solve mathematical problems.

Method notes

These appear alongside some of the examples to give more detailed help and advice about working out the answers.

Essential notes

These are other ideas which you will find useful or need to recall from previous study.

Exam tips

These tell you what you will be expected to do, or not to do, in the examination.

Stop and think

The stop and think sections present problems and questions to help you reflect on what you have just been reading. They are not straightforward practice questions - you have to think carefully to answer them!

Practice examination section

At the end of the book you will find a section of practice examination questions which help you prepare for the ones in the examination itself. Answers with full workings out are provided so that you can see exactly where you are getting things wrong or right!

Notation and formulae

The notation and formulae used in this examination module are listed at the end of the book just before the index for easy reference. The formulae list shows both what you need to know and what you will be given in the exam.

Contents

Contents

Essential notes

'Indices' is the plural of the word 'index'. Indices can also be called powers.

Essential notes

'Indices' is the plural of the word 'index'. Indices can also be called powers.

Essential notes

$a^{\frac{1}{m}} = \sqrt[m]{a}$ is called the mth root of a.

Exam tips

These rules need to be learnt. They are not in the Formula Booklet and will occur often in A Level Mathematics

Laws of indices

The laws of **indices** allow you to manipulate and simplify algebraic expressions.

The rules for manipulating **powers** or indices are:

1. $a^m \times a^n = a^{m+n}$

2. $a^m \div a^n = a^{m-n}$

3. $(a^m)^n = a^{mn}$

4. $a^{-m} = \dfrac{1}{a^m}$

5. $a^{\frac{1}{m}} = \sqrt[m]{a}$

6. $a^{\frac{n}{m}} = \sqrt[m]{a^n}$

7. $a^0 = 1$

In the rules above a is a positive number called the **base** number.

Evaluating expressions using the laws of indices

Example

Evaluate the following without using a calculator.

a) 2^{-4} b) $49^{\frac{1}{2}}$ c) $49^{\frac{3}{2}}$ d) $\sqrt{6\frac{1}{4}}$

e) $\sqrt[3]{3\frac{3}{8}}$ f) $\left(\dfrac{9}{16}\right)^{-\frac{1}{2}}$

Answer

a) $2^{-4} = \dfrac{1}{2^4} = \dfrac{1}{16}$ Rule 4

b) $49^{\frac{1}{2}} = \sqrt{49} = 7$ Rule 5

c) $49^{\frac{3}{2}} = \sqrt{49^3} = (\sqrt{49})^3 = 7^3 = 343$ Rule 6

d) $\sqrt{6\frac{1}{4}} = \sqrt{\dfrac{25}{4}} = \dfrac{\sqrt{25}}{\sqrt{4}} = \dfrac{5}{2}$

e) $\sqrt[3]{3\frac{3}{8}} = \sqrt[3]{\left(\dfrac{27}{8}\right)} = \dfrac{\sqrt[3]{27}}{\sqrt[3]{8}} = \dfrac{3}{2}$

f) $\left(\dfrac{9}{16}\right)^{-\frac{1}{2}} = \dfrac{1}{\left(\dfrac{9}{16}\right)^{\frac{1}{2}}} = \dfrac{1}{\sqrt{\dfrac{9}{16}}} = \dfrac{1}{\dfrac{3}{4}} = \dfrac{4}{3}$ Rules 4 and 5

Method notes

In d) and e) write $6\frac{1}{4}$ and $3\frac{3}{8}$ as improper fractions. Then take the roots of the numerator and denominator separately.

Simplifying expressions using the laws of indices

Example

Simplify the following expressions

a) $(7x^3y^4)^2$ b) $\left(\dfrac{5x}{2y^2}\right)^2$ c) $\sqrt[3]{125x^3y^6}$

Answer

a) $(7x^3y^4)^2 = (7x^3y^4) \times (7x^3y^4) = 49\,x^6y^8$

b) $\left(\dfrac{5x}{2y^2}\right)^2 = \left(\dfrac{5x}{2y^2}\right) \times \left(\dfrac{5x}{2y^2}\right) = \dfrac{25x^2}{4y^4}$

c) $\sqrt[3]{125x^3y^6} = (125x^3y^6)^{\frac{1}{3}} = 125^{\frac{1}{3}} \times (x^3)^{\frac{1}{3}} \times (y^6)^{\frac{1}{3}} = 5xy^2$

Essential notes

In a) and b) we see that, to square any algebraic expression of numbers and powers, we just square the numbers and multiply the power of each variable by 2

Change of base

Some expressions may involve different bases. Care is needed when simplifying these expressions. The technique is to look for a common base number in each part of the expression.

Example

Simplify $9^5 \times 27^{-2}$

Answer

$9^5 = (3^2)^5 = 3^{10}$

$27^{-2} = \dfrac{1}{27^2} = \dfrac{1}{(3^3)^2} = \dfrac{1}{3^6}$

$9^5 \times 27^{-2} = 3^{10} \times \dfrac{1}{3^6} = 3^4$

Method notes

In the given expression 9 and 27 can be written as powers of 3 which will be our common base. So we now convert 9^5 and 27^{-2} to powers of the common base 3

Surds

When taking square roots, some can be evaluated as integers such as $\sqrt{25} = 5$ or $\sqrt{121} = 11$. This is always true for the square root of square numbers.

Other square roots, such as $\sqrt{2}$, $\sqrt{5}$ are examples of irrational numbers. They cannot be evaluated as integers since 2 and 5 are not square numbers.

In 820 AD the Persian mathematician al-khwarizmi (after whom algorithm is named) called irrational numbers 'inaudible' which was translated into the Latin 'surdus' (deaf or mute).

Hence an irrational number written as the square root of a prime number became known as a **surd**.

$\sqrt{2}$ is the exact value of the square root of 2. If you try to find its value on a calculator you will only get an approximation.

It is possible to manipulate surds following the usual rules of arithmetic.

Essential notes

Numbers can be classified in two ways. A **rational number** is a number that can be written as a fraction of two integers. Numbers which are not rational are called **irrational numbers**. $\sqrt{2}$ and π are examples of irrational numbers.

The square root symbol $\sqrt{}$ always gives a positive answer.

Essential notes

Each of these rules involving the square root symbol is only true for positive numbers.

Rules for surds

$$\sqrt{a \times b} = \sqrt{a} \times \sqrt{b}$$

$$\sqrt{\frac{a}{b}} = \frac{\sqrt{a}}{\sqrt{b}}$$

Using surds

Surds occur naturally when using Pythagoras' theorem for right-angled triangles. For example, the hypotenuse h in Figure 1.1 is found by using

$$h^2 = 2^2 + 5^2 \Rightarrow h = \sqrt{29}$$

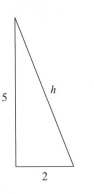

Fig. 1.1
Right-angled triangle.

Simplifying square roots

It is usual to express square roots in their simplest form involving surds. For example,

$\sqrt{20} = \sqrt{4 \times 5} = \sqrt{4} \times \sqrt{5}$ which would be written as $2\sqrt{5}$. The number inside the square root sign cannot be simplified further so the answer $2\sqrt{5}$ is now in its simplest form.

Example
Simplify the following surds.

a) $\sqrt{44}$ b) $\sqrt{640}$

Answer
a) $\sqrt{44} = \sqrt{4 \times 11} = \sqrt{4} \times \sqrt{11} = 2\sqrt{11}$
b) $\sqrt{640} = \sqrt{4 \times 16 \times 10} = \sqrt{4} \times \sqrt{16} \times \sqrt{10}$
$$= 2 \times 4 \times \sqrt{10} = 8 \times \sqrt{2 \times 5} = 8\sqrt{2}\sqrt{5}$$

Simplifying expressions using surds

Any expression involving surds is in its simplest form if:

• numbers inside the square roots are prime numbers

• like terms are gathered together

• fractions do not contain surds in their denominator.

Method notes

The first step in each of these examples is to find any square factors of the numbers given.

As in algebra with variables, we often omit the multiplication symbol between numbers and surds.

Example
Simplify the following as a single surd in its simplest form:

a) $2\sqrt{3} + 5\sqrt{3} - 3\sqrt{3}$

b) $(3\sqrt{5} - 1) \times (1 + \sqrt{5})$

c) $\sqrt{2} - \sqrt{18} + \sqrt{32}$

d) $(2 + 3\sqrt{5}) \times (2 - 3\sqrt{5})$

Answer

a) $2\sqrt{3}+5\sqrt{3}-3\sqrt{3}=4\sqrt{3}$

b) $(3\sqrt{5}-1)\times(1+\sqrt{5})=3\sqrt{5}+(3\sqrt{5}\times\sqrt{5})-1-\sqrt{5}$

$\quad=(3\sqrt{5}+15-1-\sqrt{5}=2\sqrt{5}+14$

c) $\sqrt{2}-\sqrt{18}+\sqrt{32}=\sqrt{2}-\sqrt{9\times2}+\sqrt{16\times2}$

$\quad=\sqrt{2}-3\sqrt{2}+4\sqrt{2}=2\sqrt{2}$

d) $(2+3\sqrt{5})\times(2-3\sqrt{5})=4-6\sqrt{5}+6\sqrt{5}-45=4-45=-41$

Method note

In a) all are multiples of $\sqrt{3}$ so we collect them together using arithmetic rules.

In b) expand the brackets in the usual way and then simplify by collecting together the terms involving $\sqrt{5}$

In c) write each term in terms of $\sqrt{2}$

In d) $3\sqrt{5}\times3\sqrt{5}$ means $3\times3\times\sqrt{5}\times\sqrt{5}=9\times5=45$

Rationalising the denominator

In part d) of the example the answer does not contain a surd.

If we multiply $(a+b\sqrt{c})$ by $(a-b\sqrt{c})$ we remove the surd from the answer.

$(a+b\sqrt{c})\times(a-b\sqrt{c})=a^2+ab\sqrt{c}-ab\sqrt{c}-b^2c=a^2-b^2c$

$(a-b\sqrt{c})$ is often called the **conjugate** of $(a+b\sqrt{c})$.

Rationalising the denominator means removing a surd from the denominator of a fraction. The appropriate conjugate expression is used to do this.

Example

Rationalise the denominators of the following fractions:

a) $\dfrac{3\sqrt{5}}{\sqrt{7}}$ b) $\dfrac{5}{3-\sqrt{2}}$ c) $\dfrac{3+\sqrt{5}}{4+\sqrt{3}}$

Answer

a) $\dfrac{3\sqrt{5}}{\sqrt{7}}=\dfrac{3\sqrt{5}}{\sqrt{7}}\times\dfrac{\sqrt{7}}{\sqrt{7}}=\dfrac{3\sqrt{5}\sqrt{7}}{7}$

b) $\dfrac{5}{3-\sqrt{2}}=\dfrac{5}{(3-\sqrt{2})}\times\dfrac{(3+\sqrt{2})}{(3+\sqrt{2})}=\dfrac{15+5\sqrt{2}}{9-2}=\dfrac{15+5\sqrt{2}}{7}$

c) $\dfrac{3+\sqrt{5}}{4+\sqrt{3}}=\dfrac{(3+\sqrt{5})}{(4+\sqrt{3})}\times\dfrac{(4-\sqrt{3})}{(4-\sqrt{3})}=\dfrac{(3+\sqrt{5})(4-\sqrt{3})}{16-3}$

$\quad=\dfrac{(3+\sqrt{5})(4-\sqrt{3})}{13}$

Method note

In a) multiply the numerator and denominator by $\sqrt{7}$

In b) the denominator is $3-\sqrt{2}$ so multiply the numerator and denominator by the **conjugate**, $3+\sqrt{2}$

In c) the denominator is $4+\sqrt{3}$ so multiply the numerator and denominator by the **conjugate**, $4-\sqrt{3}$

Stop and think 1

Is it ever true that $\sqrt{a+b}=a+b$?

Is it ever true that $\sqrt{a-b}=a-b$?

Quadratic functions

The product of two linear functions such as f$(x) = x - 2$ and g$(x) = x - 3$ introduces a new function $(x - 2)(x - 3) = x^2 - 5x + 6$

The highest power is 2 so it is called a **quadratic function**.

Any expression with a term in x^2, with or without a term in x and with or without a constant, is a quadratic function. The following are all quadratic functions in a variable x:

$y = x^2$

$y = 3x^2 + 11$

$y = 2x^2 - 3x + 5$

You can use other letters to represent the variables, for example,

$s = 14t - 10t^2$

is a quadratic function in t.

The general form of a quadratic function in the variable x is

$y = ax^2 + bx + c$

where a, b and c are constants, are called **coefficients** and $a \neq 0$

- a is the coefficient of x^2
- b is the coefficient of x^1
- c is the constant term (or coefficient of x^0)

Graphs of quadratic functions

The graph of a quadratic function is called a parabola. Every parabola is symmetrical about a line of symmetry through the vertex as shown in Figure 1.2.

Fig. 1.2
Shapes of parabolas.

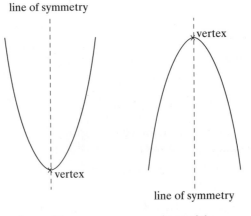

line of symmetry

vertex

vertex

line of symmetry

shape of the parabola for $a > 0$

shape of the parabola for $a < 0$

Example

A quadratic function is defined by $y = x^2 - 2x - 3$

a) Plot a graph of the function.

b) Find the coordinates of the y-intercept.

c) Find the coordinates of the points where the graph intercepts the x-axis.

d) Draw in, and give the equation of, the line of symmetry.

e) Write down the coordinates of the vertex.

Answer

A table of values for $y = x^2 - 2x - 3$ gives:

x	−3	−2	−1	0	1	2	3	4	5
y	12	5	0	−3	−4	−3	0	5	12

a)

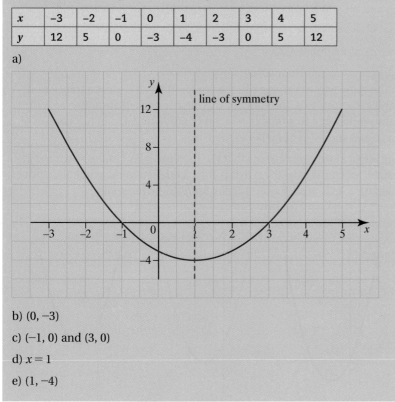

Fig. 1.3
Graph of quadratic function $y = x^2 - 2x - 3$.

b) $(0, -3)$

c) $(-1, 0)$ and $(3, 0)$

d) $x = 1$

e) $(1, -4)$

The position and shape of the parabolic graph for a quadratic function $y = ax^2 + bx + c$ depends on the values of a, b and c.

The constant c plays a similar role as it does for the straight line graph $y = mx + c$. It gives the coordinates of the y-intercept $(0, c)$.

The constant a gives either a concave up shape for $a > 0$ ('a smiley face') or a concave down shape ('a sad face') for $a < 0$ as shown in Figure 1.2. The effect of varying a is to stretch the parabola parallel to the y-axis.

The effect of varying b is less easy to interpret. Figure 1.4 shows the graphs of three quadratic functions:

Fig. 1.4
The effect of varying b.

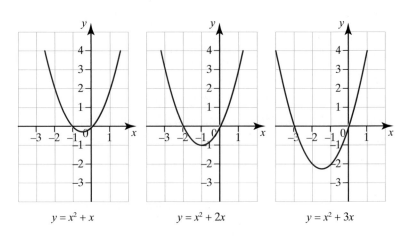

$$y = x^2 + x \qquad y = x^2 + 2x \qquad y = x^2 + 3x$$

Exam tips

You need to learn the effect of varying b in the equation $y = x^2 + bx$.

In the same way as varying c in the equation $y = x^2 + c$ is a translation parallel to the y-axis, the effect of varying b in the equation $y = x^2 + bx$ is a translation, but this time in two directions: a translation of $\frac{b}{2}$ parallel to the x-axis and $-\left(\frac{b}{2}\right)^2$ parallel to the y-axis.

The effect of varying c means that the parabola may cut the x-axis in two points, touch the axis (the x-axis is then a tangent to the parabola) or not meet the x-axis at all. Figure 1.5 illustrates these three cases.

Fig. 1.5
The effect of varying c.

Essential notes

The points where the graph cuts or touches the x-axis are called the **roots** of the quadratic equation $ax^2 + bx + c = 0$

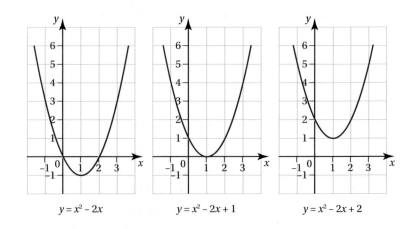

$$y = x^2 - 2x \qquad y = x^2 - 2x + 1 \qquad y = x^2 - 2x + 2$$

Completing the square

To understand fully the properties of the more general quadratic function $y = ax^2 + bx + c$ and its graph, we need to transform the equation of the function into a different form using the method of **completing the square**.

We aim to write $y = ax^2 + bx + c$ in the form $y = p(x + q)^2 + r$ which then compares easily with $y = x^2 + d$.

Example

Write the quadratic function $x^2 + 6x + 11$ in the form $(x + q)^2 + r$. In this case the coefficient of x^2 is 1

Answer

Step 1: Let $x^2 + 6x + 11 = (x + q)^2 + r$

Step 2: Expand the bracket so that:

$x^2 + 6x + 11 = x^2 + 2qx + q^2 + r$

Step 3: Check that the coefficients of x^2 are both equal (here both $= 1$)

Step 4: Compare coefficients of x in step 2: $\quad 6 = 2q \quad \Rightarrow \quad 3 = q$

Step 5: Compare constant terms in step 2: $\quad 11 = q^2 + r$

But from step 4: $q = 3$ so $11 = 3^2 + r$ and $r = 2$

Putting these number values of q and r into step 1 means that

$x^2 + 6x + 11 = (x + 3)^2 + 2$

Essential notes

The expansion of $(a \pm b)^2$ occurs often in A Level Mathematics and the following formulas are worth learning:

$(a + b)^2 = a^2 + 2ab + b^2$
$(a - b)^2 = a^2 - 2ab + b^2$

Note the $2ab$ in each expression.

This process of rewriting the quadratic function is known as **completing the square**.

Example

Write the quadratic function $2x^2 + 12x + 3$ in the form $2(x + q)^2 + r$. In this case the coefficient of x^2 is not 1

Answer

Step 1: Let $2x^2 + 12x + 3 = 2(x + q)^2 + r$

Step 2: Expand the squared bracket
$2x^2 + 12x + 3 = 2(x^2 + 2qx + q^2) + r$

Step 3: Multiply out giving $2x^2 + 12x + 3 = 2x^2 + 4qx + 2q^2 + r$

Step 4: Check the coefficients in step 3 of x^2 are equal (here both $= 2$)

Step 5: Compare x coefficients in step 3: $\quad 12 = 4q \quad \Rightarrow \quad 3 = q$

Step 6: Compare constant terms in step 3: $\quad 3 = 2q^2 + r$

$\qquad$ but, from step 5, $q = 3$ $\quad$ giving $\qquad 3 = 2(3)^2 + r$

$\qquad\qquad\qquad\qquad\qquad\qquad\qquad\qquad 3 = 18 + r$

$\qquad\qquad\qquad$ so $\qquad\qquad\qquad\qquad r = -15$

Using these number values of q and r in step 1 means that

$2x^2 + 12x + 3 = 2(x + 3)^2 - 15$ and we have completed the square.

Essential notes

The value of q is half the coefficient of x (in the given quadratic formula) in this case $\frac{6}{2} = 3$

The value of r is the constant term (in the given quadratic formula) $c - q^2$ is in this case $11 - 3^2$ so $r = 2$

Stop and think 2

Is it always true that, if x is an integer, $x^2 + 4x + 4$ gives a square number whatever the value of x?

Discriminant

If we wish to write the general quadratic function $ax^2 + bx + c$ in the completed square form $p(x + q)^2 + r$, the method is as follows:

Step 1: Take out a as a common factor of the first two terms

$$ax^2 + bx + c \equiv a(x^2 + \frac{b}{a}x) + c$$

Step 2: Complete the square by halving the coefficient of x.

$$a\left(x^2 + \frac{b}{a}x\right) + c = a\left[\left(x + \frac{b}{2a}\right)^2 - \frac{b^2}{4a^2}\right] + c = a\left(x + \frac{b}{2a}\right)^2 - \frac{b^2}{4a} + c$$

$$= a\left(x + \frac{b}{2a}\right)^2 - \left[\frac{b^2 - 4ac}{4a}\right]$$

Step 3: $ax^2 + bx + c = a\left(x + \frac{b}{2a}\right)^2 - \left(\frac{b^2 - 4ac}{4a}\right)$

Now we can complete the graphs of a quadratic function.

The different cases depend on the value of $b^2 - 4ac$ called the **discriminant** of the quadratic function.

The six possible cases are shown in Figure 1.6.

Fig. 1.6
Six cases for the discriminant.

Essential notes

The discriminant is used to determine the number of roots of a quadratic equation.

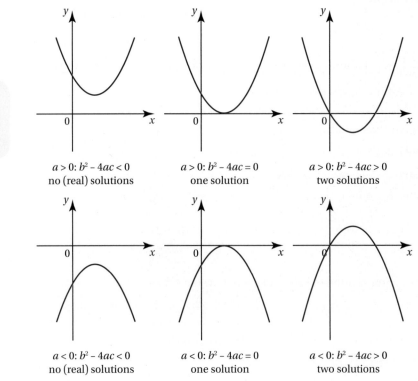

$a > 0$: $b^2 - 4ac < 0$
no (real) solutions

$a > 0$: $b^2 - 4ac = 0$
one solution

$a > 0$: $b^2 - 4ac > 0$
two solutions

$a < 0$: $b^2 - 4ac < 0$
no (real) solutions

$a < 0$: $b^2 - 4ac = 0$
one solution

$a < 0$: $b^2 - 4ac > 0$
two solutions

Example

The graph of the quadratic function with equation $y = x^2 + kx + 9$, where k is a constant, cuts the x-axis at two points. Find the set of values that k can take.

Answer

For two points of intersection the discriminant $b^2 - 4ac$ has to be positive.

In this example, $a = 1$ so $a > 0$, $b = k$ and $c = 9$

$b^2 - 4ac = k^2 - 4 \times 1 \times 9 = k^2 - 36$ which is the discriminant and has to be positive.

so $k^2 - 36 > 0$

$k^2 > 36$

$k > 6$ or $k < -6$

Method notes

Remember that the equation $k^2 = 36$ has two solutions $k = +6$ and $k = -6$. Numbers less than -6 will also satisfy the inequality $k^2 > 36$

Sketching graphs of quadratic functions

A sketch of a graph does not require an accurate plot using a table of values. The important features of a graph can be found algebraically and shown on a sketch graph.

Example

a) Find the coordinates where the parabola $y = 5 + 9x - 2x^2$ cuts the x-axis and the y-axis.

b) Write down the equation of the line of symmetry.

c) Sketch the graph of $y = 5 + 9x - 2x^2$

Answer

a) $y = 5 + 9x - 2x^2$

The y-intercept is given by $x = 0 \Rightarrow y = 5$

The coordinates of the y-intercept are $(0, 5)$

The x-intercepts are given by $y = 0 \Rightarrow 5 + 9x - 2x^2 = 0$

Factorising the quadratic $\Rightarrow 5 + 9x - 2x^2 = (5 - x)(1 + 2x) = 0$

Solving for $x \Rightarrow x = 5$ or $x = -\dfrac{1}{2}$

The x-intercepts are $(5, 0)$ and $\left(-\dfrac{1}{2}, 0\right)$

b) The line of symmetry is $x = \dfrac{1}{2}\left(5 + \left(-\dfrac{1}{2}\right)\right) = 2\dfrac{1}{4}$

Method notes

To find the y-intercept put $x = 0$

To find the x-intercept put $y = 0$

See page for factorising quadratic equations.

Exam tip

Give both coordinates of the intercept points in your answers.

Method notes

The line of symmetry is half way between the x-intercept points.

☞ Continued on the next page

Fig. 1.7
Sketch of $y = 5 + 9x - 2x^2$.

c)

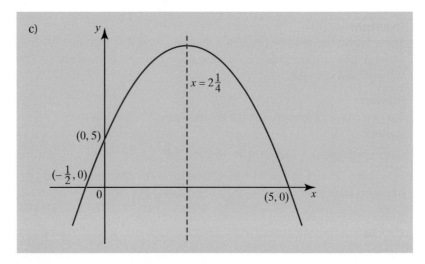

Stop and think 3

Explain why $a(a-4) = b(b-4)$ does not lead to $a = b$ or $a = b = 4$

Solutions of quadratic equations

An equation is a mathematical statement where two quantities are equal. A **quadratic equation** is a statement containing a quadratic function.

The general form of a quadratic equation is $ax^2 + bx + c = 0$ and $a \neq 0$

Solving quadratic equations by factorising

Quadratic functions can be formed by multiplying two linear functions together. For example $f(x) = x - 2$ and $g(x) = x + 5$

$(x - 2) \times (x + 5) = x^2 + 3x - 10$

If we are given the quadratic equation $x^2 + 3x - 10 = 0$ to solve, we would **factorise** it by finding the two linear functions and then the solution would be quite straight forward.

Replace $x^2 + 3x - 10$ by $(x - 2) \times (x + 5)$ then solve $(x - 2) \times (x + 5) = 0$

Using the fact that if $A \times B = 0$ then either $A = 0$ or $B = 0$

$(x - 2) \times (x + 5) = 0 \Rightarrow (x - 2) = 0$ or $(x + 5) = 0$

$\Rightarrow \quad x = 2$ or $x = -5$ are the solutions

If a given quadratic function has two linear factors that can be found at a glance, this provides a method of solving quadratic equations. If you find that factorising quadratics is very difficult then there are two alternative strategies which are given below.

Essential notes

$x + 5$ is a linear function as the power of x is 1

$x - 2$ is also a linear function.

Essential notes

Always check that the solutions make the original equation true.

$x = 2$ in $x^2 + 3x - 10$ gives $2^2 + 3(2) - 10$ which is 0

Example

Solve the quadratic equation $2x^2 - 11x + 5 = 0$ by factorising.

Answer

$2x^2 - 11x + 5 = 0 \Rightarrow (2x - 1)(x - 5) = 0$

$\Rightarrow \quad (2x - 1) = 0 \text{ or } (x - 5) = 0$

$\Rightarrow \quad x = \dfrac{1}{2} \text{ or } x = 5 \text{ are the solutions}$

Solving quadratic equations by completing the square

Example

Solve the quadratic equation $x^2 - 8x - 11 = 0$ by completing the square.

Answer

$x^2 - 8x - 11 = 0$

Complete the square $\Rightarrow (x - 4)^2 - 16 - 11 = 0$

$(x - 4)^2 - 27 = 0$

Add 27 to both sides $\Rightarrow (x - 4)^2 = 27$

Take the square root of both sides: $(x - 4) = \sqrt{27} = \pm 3\sqrt{3}$

Rearrange: $x = 4 \pm 3\sqrt{3}$

Method notes

To complete the square from earlier section: halve the coefficient of x, -8 to give $q = -4$ and $r = \text{constant} - q^2$ so $r = -11 - 16$ giving $r = -27$

Example

Solve the quadratic equation $2x^2 - 12x + 17 = 0$

Answer

Take out factor of 2: $2(x^2 - 6x) + 17 = 0$ and complete the square

$2x^2 - 12x + 17 = 0 \Rightarrow 2(x - 3)^2 - 18 + 17 = 0$

$\Rightarrow 2x^2 - 12x + 17 = 0 \Rightarrow 2(x - 3)^2 - 1 = 0$

Add 1 to both sides $\Rightarrow \qquad 2(x - 3)^2 = 1$

Divide both sides by 2 $\Rightarrow \quad (x - 3)^2 = \dfrac{1}{2}$

Take the square root of both sides: $\quad \Rightarrow \quad (x - 3) = \pm \sqrt{\tfrac{1}{2}}$

Rearrange: $\qquad\qquad\qquad \Rightarrow \quad x = 3 \pm \sqrt{\tfrac{1}{2}}$ and these are the exact solutions.

Exam Tip

These are the exact answers as they include surds. In C1 you do not have use of a calculator. You are expected to give exact answers.

Method notes

Completing the square (Step 1) gives:

$ax^2 + bx + c =$

$a\left[\left(x^2 + \dfrac{b}{a}x\right) + \dfrac{c}{a}\right]$

$= a\left[\left(x + \dfrac{b}{2a}\right)^2 - \left(\dfrac{b}{2a}\right)^2 + \dfrac{c}{a}\right]$

and $-\left(\dfrac{b^2}{4a^2} - \dfrac{c}{a}\right) =$

$-\left(\dfrac{b^2 - 4ac}{4a^2}\right)$

Solving quadratic equations by using the formula

More generally if we want to solve $ax^2 + bx + c = 0$ we can use a formula. This is found like this.

Step 1: $ax^2 + bx + c = a\left(x + \dfrac{b}{2a}\right)^2 - \left(\dfrac{b^2 - 4ac}{4a}\right) = 0$

Step 2: Rearrange $\Rightarrow a\left(x + \dfrac{b}{2a}\right)^2 = \left(\dfrac{b^2 - 4ac}{4a}\right)$

Step 3: Divide both sides by $a \Rightarrow \left(x + \dfrac{b}{2a} \right)^2 = \left(\dfrac{b^2 - 4ac}{4a^2} \right)$

Step 4: Take square root of both sides $\Rightarrow \left(x + \dfrac{b}{2a} \right) = \pm \left(\dfrac{\sqrt{b^2 - 4ac}}{2a} \right)$

Step 5: Rearrange $\Rightarrow x = -\dfrac{b}{2a} \pm \left(\dfrac{\sqrt{b^2 - 4ac}}{2a} \right)$

Step 6: Rewrite $\Rightarrow x = \dfrac{-b \pm \sqrt{b^2 - 4ac}}{2a}$

This gives a general formula in terms of the coefficients of the original equation and can therefore be used to solve any quadratic equation. The sign of the discriminant $b^2 - 4ac$ tells us whether we have two solutions, one solution or no (real) solution (see page 14).

Example

Solve the quadratic equation $3x^2 - 13x - 10 = 0$ by using the formula.

Answer

$a = 3$, $b = -13$ and $c = -10$

$$x = \frac{-b \pm \sqrt{b^2 - 4ac}}{2a} = \frac{-(-13) \pm \sqrt{(-13)^2 - 4 \times 3 \times (-10)}}{2 \times 3}$$

$$x = \frac{13 \pm \sqrt{169 + 120}}{6} = \frac{13 \pm \sqrt{289}}{6} = \frac{13 \pm 17}{6}$$

so $x = 5$ or $x = -\frac{2}{3}$ are the solutions.

Stop and think 4

Without solving the equations state how many solutions there will be for each equation.

a) $2a^2 + 4a = 7$

b) $3b^2 + 1 = 0$

Simultaneous equations

Simultaneous equations are two or more equations which involve two or more variables whose values are to be found. There needs to be the same number of equations as there are unknown variables.

In A Level Mathematics you will only be asked to solve two simultaneous equations with two unknowns.

There are two algebraic methods that are commonly used.

Method of solution by substitution

In this method one of the equations is rearranged to make one variable the subject of the equation. This variable is then substituted into the second equation.

Example

Solve the pair of linear simultaneous equations:

$5x + 4y = 11$

$3x + y = 1$

Answer

$5x + 4y = 11$ (1)

$3x + y = 1$ (2)

Make y the subject of equation (2)

$y = 1 - 3x$ (3) Label new equation.

Substitute this expression for y into equation (1).

$5x + 4(1 - 3x) = 11$

$5x + 4 - 12x = 11$

$-7x = 7$

$x = -1$

Substitute this value for x into equation (3).

$y = 1 - 3(-1) = 4$

The solution is $x = -1$, $y = 4$ (both variables must have a value).

Method notes

Label each equation, for example, (1) as the first, (2) as the second for easy reference.

The subject of the equation is a variable written completely in terms of the other variable. Hence $y = 1 - 3x$ means y is the subject.

Exam tips

Use statements to explain what you are doing.

Check your answers by substituting the x and y values into **both** of the original equations to confirm that they are true.

Method of solution by elimination

In this method, one (or both) equations is multiplied by a constant to make the coefficients of one of the variables (either x or y) equal. Then, by adding or subtracting the equations, one of the variables is eliminated.

Example

Solve the pair of simultaneous equations:

$5x + 4y = 11$

$3x + y = 1$

Answer

$5x + 4y = 11$ (1)

$3x + y = 1$ (2)

Make the coefficient of y the same in both equations.

Method notes

You can choose to eliminate either x or y. Look at the coefficients and decide which is easier to eliminate. In this example choosing y involves only one multiplication by a constant.

Continued on the next page

$(2) \times 4 \Rightarrow 12x + 4y = 4$ (3) Label new equation.

$(1) \Rightarrow 5x + 4y = 11$ (1)

Eliminate y.

$(3) - (1)$ $7x + 0y = -7$

$\qquad\qquad x = -1$

Substitute for x into equation (2).

$$3(-1) + y = 1 \Rightarrow y = 4$$

The solution is $x = -1$, $y = 4$

Solution of one linear equation with one quadratic equation

For a pair of simultaneous equations where one is a linear equation and the other a quadratic equation the method of substitution is usually used. The linear equation is rearranged to make one variable the subject. This variable is then substituted into the quadratic equation.

Answers can be two pairs of solutions for x and y, one pair of solutions for x and y or no solutions. The answer will depend on the quadratic equation which occurs after the substitution step.

Essential notes

The most common application of this topic is the intersection of a straight line with a parabola.

Method notes

You can always check how many solutions to expect by evaluating the discriminant. Here you expect two solutions because

$\sqrt{b^2 - 4ac}$
$= \sqrt{3^2 - 4 \times 1 \times (-4)}$
$= 5 > 0$

Exam tips

Make sure that you show the solution pairs clearly. The coordinate notation is generally used to show the x and y answers.

Example

Solve the pair of simultaneous equations:

$2x^2 - 3x - y = 1$

$y - 3x = 7$

Answer

$2x^2 - 3x - y = 1$ (1)

$y - 3x = 7$ (2)

Make y the subject of equation (2).

$y = 7 + 3x$ (3)

Substitute for y into equation (1).

$2x^2 - 3x - (7 + 3x) = 1$

$2x^2 - 3x - 7 - 3x = 1$

$2x^2 - 6x - 8 = 0$

$x^2 - 3x - 4 = 0$

$(x - 4)(x + 1) = 0$

$x = 4$ or $x = -1$

Substitute for both values of x into equation (3) to give the corresponding values of y.

$x = 4 \Rightarrow y = 7 + 3(4) = 19$ gives the solution (4, 19).

$x = -1 \Rightarrow y = 7 + 3(-1) = 4$ gives the solution (−1, 4).

Example

Solve the pair of simultaneous equations:

$2xy = 9$

$3x + 4y = 15$

Answer

Either:

$2xy = 9$ (1)

$3x + 4y = 15$ (2)

Make y the subject of equation (1).

$y = \dfrac{9}{2x}$ (3)

Substitute for y into equation (2).

$3x + 4 \times \dfrac{9}{2x} = 15$

$3x + \dfrac{18}{x} = 15$

$3x^2 - 15x + 18 = 0$

$x^2 - 5x + 6 = 0$

Or:

$2xy = 9$ (1)

$4y = 15 - 3x$ (2)

$4xy = 18$ (3)

$x(15 - 3x) = 18$

$15x - 3x^2 = 18$

$3x^2 - 15x + 18 = 0$

$x^2 - 5x + 6 = 0$

Then:

$(x - 2)(x - 3) = 0$

$x = 2$ or $x = 3$

Substitute for both values of x into equation (3) to give the corresponding values of y.

$x = 2 \Rightarrow y = \dfrac{9}{2x} = \dfrac{9}{4}$ gives the solution $\left(2, \dfrac{9}{4}\right)$.

$x = 3 \Rightarrow y = \dfrac{9}{2x} = \dfrac{9}{6} = \dfrac{3}{2}$ gives the solution $\left(3, \dfrac{3}{2}\right)$.

Linear and quadratic inequalities

The same arithmetic processes are used to solve inequalities as for solving equations, with one very important exception. When you multiply or divide an inequality by a negative number, you must reverse the inequality sign.

For example, if $-x < 4$ then multiplying both sides by (-1) leads to $x > -4$.

Solution of linear inequalities

As with linear equations, solving a linear inequality requires making x the subject of the inequality.

Example

Solve the following inequalities.

a) $14 - 3x < 5x + 6$

b) $5(x + 2) \geq 2(4 - x)$

c) $15 \leq (9 - 2x) < 31$

Answer

a) $14 - 3x < 5x + 6$

Add $3x$ to both sides: $\Rightarrow 14 < 8x + 6$

Subtract 6 from both sides: $\Rightarrow 8 < 8x \quad \Rightarrow \quad 1 < x$

It is usual to write the answer in the form $x > 1$

b) $5(x + 2) \geq 2(4 - x)$

Expand the brackets $\Rightarrow 5x + 10 \geq 8 - 2x$

Add $2x$ to both sides: $\Rightarrow 7x + 10 \geq 8$

Subtract 10 from both sides: $\Rightarrow 7x \geq -2 \Rightarrow x \geq -\frac{2}{7}$

c) $15 \leq (9 - 2x) < 31$

We deal with each inequality separately:

$15 \leq (9 - 2x) \Rightarrow 6 \leq -2x \Rightarrow 3 \leq -x \Rightarrow -3 \geq x$ i.e. $x \leq -3$

$(9 - 2x) < 31 \Rightarrow -2x < 22 \Rightarrow -x < 11 \Rightarrow x > -11$

Combining both inequalities then gives $-11 < x \leq -3$

Solution of quadratic inequalities

To solve a quadratic inequality such as $ax^2 + bx + c \geq 0$ first solve the quadratic equation $ax^2 + bx + c = 0$. Then, from a sketch graph of $y = ax^2 + bx + c$, deduce the interval(s) for which the inequality holds.

Example

Find the range of values for which $x^2 + 3x - 10 \leq 0$

Answer

First solve the equation $x^2 + 3x - 10 = 0$ to find where the graph crosses the x-axis.

$x^2 + 3x - 10 = 0 \implies (x+5)(x-2) = 0 \implies x = -5$ or $x = 2$

A sketch of the graph $y = x^2 + 3x - 10$ shows that if y has to be ≤ 0 as given in the question, then the curve has to be below the x-axis or on it.

Fig. 1.8
Sketch of the graph $y = x^2 + 3x - 10$.

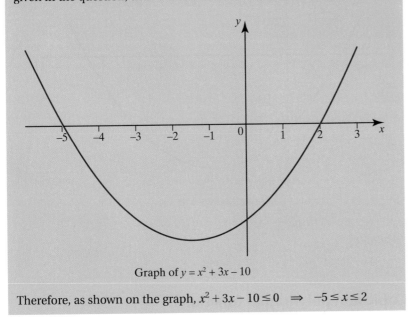

Graph of $y = x^2 + 3x - 10$

Therefore, as shown on the graph, $x^2 + 3x - 10 \leq 0 \implies -5 \leq x \leq 2$

Continued on the next page

Method notes

This is a quadratic equation in x so b is the coefficient of x giving $b = -k$

c is the constant giving $c = k + 2$

Example

Find the values of the constant k for which the equation $x^2 - kx + (k+2) = 0$ has two real roots.

Answer

For two real roots, the discriminant must be positive.

$a = 1, b = -k, c = (k+2)$

$b^2 - 4ac = k^2 - 4(k+2) > 0 \Rightarrow k^2 - 4k - 8 > 0$

Solving $k^2 - 4k - 8 = 0 \Rightarrow (k-2)^2 - 4 - 8 = 0 \Rightarrow (k-2)^2 - 12 = 0$

$\Rightarrow k = 2 \pm \sqrt{12} = 2 \pm 2\sqrt{3}$

A sketch of the graph $y = k^2 - 4k - 8$ shows that the graph crosses the k-axis where $k = 2 - 2\sqrt{3}$ and $2 + 2\sqrt{3}$ but if the discriminant $= y$ and is > 0 then this is where the graph is above the k-axis.

Fig. 1.9
Sketch of the graph $y = k^2 - 4k - 8$.

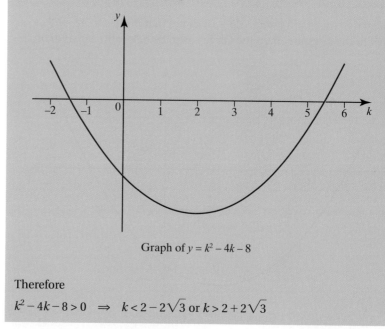

Graph of $y = k^2 - 4k - 8$

Therefore

$k^2 - 4k - 8 > 0 \Rightarrow k < 2 - 2\sqrt{3}$ or $k > 2 + 2\sqrt{3}$

Cubic and polynomial functions

The product of a linear function such as $f(x) = 2x + 3$ and a quadratic function such as $g(x) = x^2 + x - 5$ introduces a new function:

$$(2x + 3) \times (x^2 + x - 5) = 2x(x^2 + x - 5) + 3(x^2 + x - 5)$$
$$= 2x^3 + 2x^2 - 10x + 3x^2 + 3x - 15$$
$$= 2x^3 + 5x^2 - 7x - 15$$

The highest power of x is x^3.

This is an example of a **cubic function**.

The general form of a cubic function is

$y = ax^3 + bx^2 + cx + d$

The constants a, b, c and d are called **coefficients** of the cubic function and $a \neq 0$.

Consider the cubic function $2x^3 + 5x^2 - 7x - 15$

the coefficient of x^3 is 2 $a = 2$

the coefficient of x^2 is 5 $b = 5$

the coefficient of x is -7 $c = -7$

the constant term -15 can be regarded as the coefficient of x^0.

Linear, quadratic and cubic functions are example of **polynomial functions** in which each term is a multiple of a power of x.

Other polynomials which have special names are

quartic functions e.g. $x^4 + 11x^3 - 2x^2 - 5x + 3$

quintic functions e.g. $3x^5 - 5x^4 + 12x^3 + x^2 - 7x + 13$

The **degree** of a polynomial is the index of the highest power of x with a non-zero coefficient. For example, $9x^7 - 2x^5 + 12x^4 + x^2 - 17x + 3$ is a polynomial of degree 7.

Arithmetic of polynomials
Polynomials can be added, subtracted, multiplied or divided.

Adding and subtracting polynomials

> #### Example
> Two polynomials are given by $f(x) = x^3 - 3x^2 + 2x - 7$ and $g(x) = x^2 + 5x + 1$. Simplify $f(x) + g(x)$ and $f(x) - g(x)$.
>
> #### Answer
> $f(x) + g(x) = (x^3 - 3x^2 + 2x - 7) + (x^2 + 5x + 1)$
>
> $\qquad = x^3 - 2x^2 + 7x - 6$
>
> $f(x) - g(x) = (x^3 - 3x^2 + 2x - 7) - (x^2 + 5x + 1)$
>
> $\qquad = x^3 - 4x^2 - 3x - 8$
>
> The result of adding or subtracting a cubic function and a quadratic function is a cubic function.
>
> In general, if a polynomial of degree m, $f(x)$, is added to a polynomial of degree n, $g(x)$, then:
>
> $f(x) + g(x)$ is a polynomial whose degree is the greater of m and n.
>
> $f(x) - g(x)$ is a polynomial whose degree is the greater of m and n unless $m = n$.
>
> If $m = n$ the degree of $f(x) - g(x)$ could be less than m (and n). For example, if $f(x) = 4x^3 + 3x^2 - 7$ and $g(x) = 4x^3 + 5x + 1$ then $f(x) - g(x) = 3x^2 - 5x - 8$. Subtracting these two cubics results in a quadratic, because the coefficients of the highest powers of $f(x)$ and $g(x)$ are equal.

Essential notes

a is the coefficient of x^3

b is the coefficient of x^2

c is the coefficient of x^1

d is the coefficient of x^0 (or the constant term)

Essential notes

In C1 you will *not* need to divide two polynomials.

Method notes

Collect like terms (terms with the same power) together and be careful with the signs!

Method notes

List f(x) terms horizontally.

List g(x) terms vertically.

Complete the table by multiplying each horizontal term by each vertical term.

Collect all like terms together.

Multiplying polynomials

Example

Two polynomials are given by $f(x) = x^3 - 3x^2 + 2x - 7$ and $g(x) = x^2 + 5x + 1$

Simplify $f(x) \times g(x)$.

Here $f(x)$ is a cubic polynomial (degree 3) and $g(x)$ is a quadratic polynomial (degree 2)

Answer

The table contains all the terms needed in the multiplication of $f(x)$ and $g(x)$.

	x^3	$-3x^2$	$2x$	-7
x^2	x^5	$-3x^4$	$2x^3$	$-7x^2$
$5x$	$5x^4$	$-15x^3$	$10x^2$	$-35x$
1	x^3	$-3x^2$	$2x$	-7

This table shows every term in $f(x)$ is multiplied by every term in $g(x)$. Adding the 12 terms in the table gives

$f(x) \times g(x) = x^5 + 2x^4 - 12x^3 - 33x - 7$

The degree of the product $f(x) \times g(x)$ is 5 which is the sum of the degrees of $f(x)$ and $g(x)$ i.e. $3 + 2$

Essential notes

In general if a polynomial of degree m, $f(x)$, is multiplied by a polynomial of degree n, $g(x)$, then the degree of $f(x) \times g(x)$ is $m + n$.

It is not necessary to write out all the terms in a table. It is, however, important that every term in one polynomial is multiplied by every term in the second polynomial.

Example

Expand and simplify $(2x - 1)(x^2 + 4) + (x - 3)(x^3 + 1)$.

Answer

$(2x - 1)(x^2 + 4) = 2x(x^2 + 4) - (x^2 + 4) = 2x^3 + 8x - x^2 - 4$

$$= 2x^3 - x^2 + 8x - 4$$

$(x - 3)(x^3 + 1) = x(x^3 + 1) - 3(x^3 + 1) = x^4 + x - 3x^3 - 3$

$$= x^4 - 3x^3 + x - 3$$

$(2x - 1)(x^2 + 4) + (x - 3)(x^3 + 1) =$

$(2x^3 - x^2 + 8x - 4) + (x^4 - 3x^3 + x - 3) = x^4 - x^3 - x^2 + 9x - 7$

Exam tips

In C1 you will be asked to factorise a cubic polynomial with a common factor x.

Factorising polynomials

Quadratic polynomials such as $x^2 + 5x + 4$ can be factorised as the product of two linear factors $(x + 1)(x + 4)$.

A cubic polynomial can also be factorised as a product of factors. For example, $x^3 + 7x^2 + 14x + 8$ can be written as $(x + 1)(x + 2)(x + 4)$.

Example

Factorise $x^3 - 3x^2 + 4x$.

Answer

It is easy to see that the cubic polynomial $x^3 - 3x^2 - 4x$ has a factor of x.

$$x^3 - 3x^2 - 4x = x(x^2 - 3x - 4) = x(x + 1)(x - 4)$$

Method notes

Remove the factor x. Then factorise the quadratic.

Stop and think 5

State a value of x which is a solution of the equation $x^3 - 1 = 0$

How many other solutions are there?

Graphs of quadratic functions

Sketching graphs of quadratic functions

Figure 1.10 shows the graph of the quadratic function $y = x^2 - 6x + 8$.

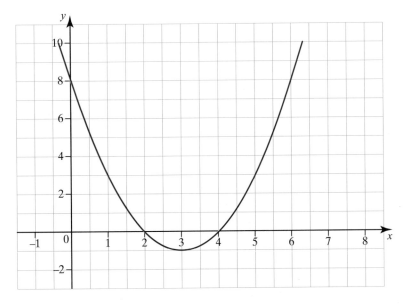

Method notes

Complete a table of values for x and y giving the following

$x = 0$ $y = 8$

$x = 2$ $y = 0$

$x = 4$ $y = 0$

$x = 6$ $y = 8$

Fig.1.10
Graph $y = x^2 - 6x + 8$.

Essential notes

When sketching quadratics you have seen two types of turning points.

For a (local) **maximum** the slope of the graph (gradient of tangent to curve) changes from positive to negative.

For a (local) **minimum** the slope of a graph (gradient of tangent to curve) changes from negative to positive.

The graph shows the essential features of a quadratic function. In this example there is a minimum value at the point $(3, -1)$.

From the table of values if $x = 0$ then $y = 8$ giving the y-intercept as $(0, 8)$.

The roots of the quadratic equation are found when $x^2 - 6x + 8 = 0$ so from the graph this is where $y = 0$ or where the graph crosses the x-axis. In this example it is when $x = 2$ and $x = 4$

This means that the function $y = x^2 - 6x + 8$ can be factorised as $(x - 2)(x - 4)$.

Therefore where $(x - 2)(x - 4) = 0$ so $x^2 - 6x + 8 = 0$ and vice versa. Roots of equations can be also be found algebraically by factorising.

A graph of a function shows the important features of the function. Usually all that is required is a sketch graph, rather than a plot of a graph from a table of values. The following tips give a strategy for graph sketching:

- Draw and label the axes.
- Find the y-intercept(s) by putting $x = 0$ into the equation and show this point on your graph.
- Find the x-intercept(s) by putting $y = 0$ into the equation and show these points on your graph. The x-values are called the roots of the equation $y = 0$.
- Find what happens to y when x is large and positive (written as $\lim_{x \to \infty}$ where lim stands for the limit of a function).
- Find what happens to y when x is large and negative (written as $\lim_{x \to -\infty}$).
- Draw on the graph any asymptotes - lines along which the curve approaches infinity. This means that as x and y increase in value they begin to get nearer and nearer to a line (the asymptote) but never reach it.

Graphs of cubic functions

Sketching graphs of cubic functions

Essential notes

Here 'lim' stands for the limit of the function y.

In C1 any asymptotes will be horizontal or vertical.

Method notes

$y = 0$ in the equation gives the x- intercept.

$x = 0$ in the equation gives the y-intercept.

Let $x \to \infty$ in $y = x^3$ so y is $\lim_{x \to \infty} x^3 = \infty$

Let $x \to -\infty$ in $y = x^3$ so y is $\lim_{x \to -\infty} x^3 = -\infty$

Example

Sketch the graph of the equation $y = x^3$.

Answer

Step 1: put $y = 0$ and find x: $y = 0^3 = 0$ so x-intercept $= 0$

Step 2: put $x = 0$ and find y: $y = x^3 = 0^3 = 0$ so y-intercept $= 0$

So we now know that $(0, 0)$ is a point on the curve so the graph crosses both axes at $(0, 0)$.

Step 3: If x has any positive value, then y is positive.

Step 4: As x approaches infinity (∞) then y approaches infinity (∞).

Step 5: If x has any negative value, then y is negative.

Step 6: As x approaches negative infinity ($-\infty$) then y approaches negative infinity ($-\infty$).

Fig. 1.11
Sketch graph of $y = x^3$.

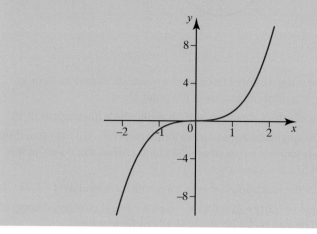

The graph in Figure 1.11 is flat at the origin (0, 0). This is a third type of turning point and is called a **point of inflexion.** The slope of the graph is positive either side of (0, 0).

Figure 1.12 shows the graph of the function $y=-x^3$.

The graph is a reflection in the x-axis of the graph of the function $y=x^3$.

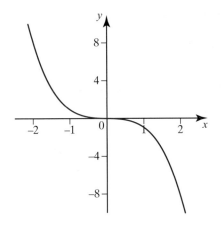

Fig. 1.12
Graph of $y=-x^3$.

Example

Sketch the graphs of the following cubic functions:

a) $y=(x+1)(x^2-3x+2)$

b) $y=(x+1)(x^2-1)$

c) $y=(x+1)(x^2+x+2)$

d) $y=(x+1)(2x-x^2)$

Continued on the next four pages

Method notes

Factorise the quadratic
$(x^2 - 3x + 2)$.

$\Rightarrow (x+1)(x-1)(x-2) = 0$

so $x = -1$, 1, or 2 when $y = 0$

$\Rightarrow$ the graph passes through
$(-1, 0)$, $(1, 0)$, $(2, 0)$.

Answer

a) $y = (x+1)(x^2 - 3x + 2)$

$x = 0 \Rightarrow y = 2$: the graph passes through $(0, 2)$.

$y = 0 \Rightarrow (x+1)(x^2 - 3x + 2) = 0$

For very large values of x the dominant term is x^3 so $y \approx x^3$ and hence as x approaches infinity (∞) then y approaches infinity (∞).

Similarly as x approaches negative infinity ($-\infty$) then y approaches negative infinity ($-\infty$).

We now have enough information to sketch a graph as in Figure 1.13.

Fig. 1.13
Graph of $y = (x+1)(x^2 - 3x + 2)$.

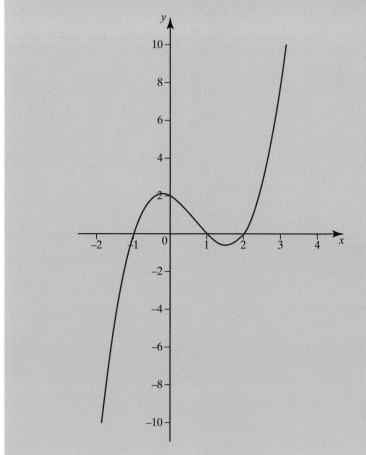

b) $y = (x + 1)(x^2 - 1)$

$\qquad x = 0 \Rightarrow y = -1$: the graph passes through $(0, -1)$.

$\qquad y = 0 \Rightarrow (x + 1)(x^2 - 1) = 0$

$\qquad\qquad \Rightarrow (x + 1)(x - 1)(x + 1) = (x + 1)^2(x - 1) = 0$

$\Rightarrow$ the graph passes through $(-1, 0)$, $(1, 0)$.

As x approaches infinity (∞) then y approaches infinity (∞).

As x approaches negative infinity ($-\infty$) then y approaches negative infinity ($-\infty$).

We now have enough information to sketch a graph as in Figure 1.14.

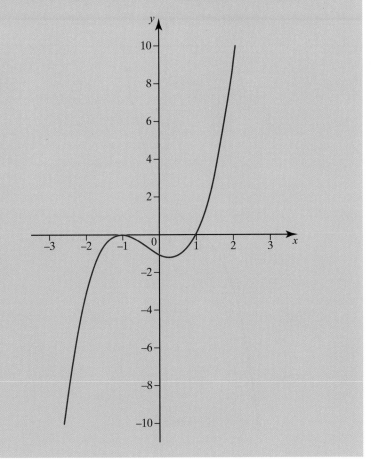

Essential notes

The squared term $(x + 1)^2$ indicates a turning point that 'touches' the x-axis. This shows in our answer by a repeated root $x = -1$ which occurs twice, meaning the x-axis is a tangent to the curve.

Fig. 1.14
Graph of $y = (x + 1)(x^2 - 1)$.

Method notes

The quadratic $(x^2 + x + 2)$ has no (real) factors because the discriminant $b^2 - 4ac = 1^2 - 4 \times 1 \times 2 = -7$

c) $y = (x+1)(x^2 + x + 2)$

$x = 0 \Rightarrow y = 2$: the graph passes through $(0, 2)$.

$y = 0 \Rightarrow (x+1)(x^2 + x + 2) = 0$

so $x = -1$ is the only 'real' solution

$\Rightarrow$ the graph passes through $(-1, 0)$.

As x approaches infinity (∞) then y approaches infinity (∞).

As x approaches negative infinity ($-\infty$) then y approaches negative infinity ($-\infty$).

We now have enough information to sketch a graph as in Figure 1.15.

Fig. 1.15
Graph of $y = (x+1)(x^2 + x + 2)$.

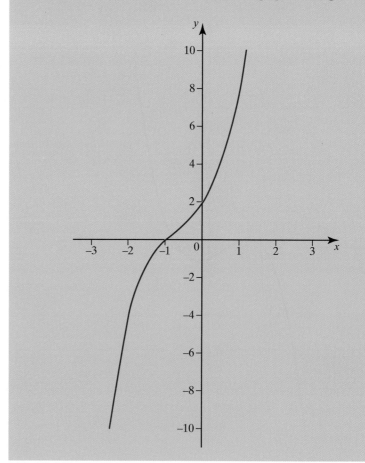

d) $y = (x+1)(2x - x^2)$

$x = 0 \Rightarrow y = 0$: the graph passes through $(0, 0)$.

$y = 0 \Rightarrow (x+1)(2x - x^2) = (x+1)x(2-x) = 0$

$\Rightarrow$ the graph passes through $(-1, 0)$, $(0, 0)$, $(2, 0)$.

As x approaches infinity (∞) then y approaches negative infinity ($-\infty$).

As x approaches negative infinity ($-\infty$) then y approaches infinity (∞).

We now have enough information to sketch a graph as in Figure 1.16.

Method notes

$y = (x+1)(2x - x^2)$

$y = 2x^2 - x^3 + 2x - x^2$

$y = -x^3 + x^2 + 2x$

$y = x(-x^2 + x + 2)$

as $x \to +\infty$ $y \to -\infty$

as $x \to -\infty$ $y \to +\infty$

Fig. 1.16
Graph of $y = (x+1)(2x - x^2)$.

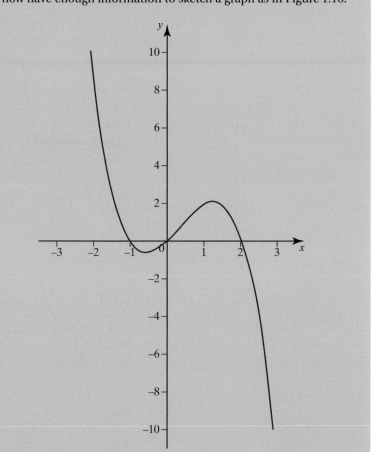

The defining features of the graphs of cubic functions are shown in the table on the following page.

Essential notes

There is always at least one root of a cubic equation.

Where the x-axis is a tangent to the graph, the root is often called a 'double root' or a 'repeated root'.

This shape of the final graph is a reflection in the x-axis because the coefficient of x^3 is negative.

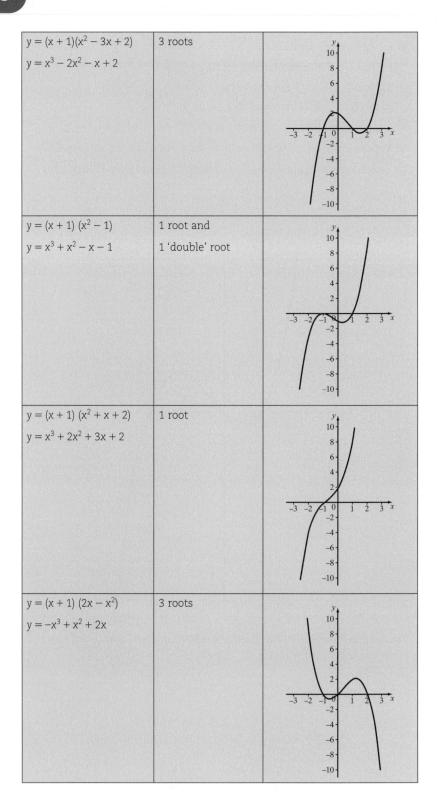

$y = (x + 1)(x^2 - 3x + 2)$ $y = x^3 - 2x^2 - x + 2$	3 roots
$y = (x + 1)(x^2 - 1)$ $y = x^3 + x^2 - x - 1$	1 root and 1 'double' root
$y = (x + 1)(x^2 + x + 2)$ $y = x^3 + 2x^2 + 3x + 2$	1 root
$y = (x + 1)(2x - x^2)$ $y = -x^3 + x^2 + 2x$	3 roots

Graphs of reciprocal functions

Sketching graphs of the functions $y = \dfrac{1}{x}$ and $y = -\dfrac{1}{x}$

Example

Sketch the graph of the reciprocal function $y = \dfrac{1}{x}$.

Answer

If x has any positive value, then y is positive. As x approaches 0 then y approaches infinity (∞).

As x approaches infinity (∞) then y tends towards the x-axis from above.

If x has any negative value, then y is negative. As x approaches 0 then y approaches minus infinity ($-\infty$).

As x approaches negative infinity ($-\infty$) then y tends towards the x-axis from below.

There are no x-intercepts and y-intercepts for $y = \dfrac{1}{x}$. The functions are undefined for $x = 0$ and for $y = 0$.

Essential notes

The graph of $y = -\dfrac{1}{x}$ is a reflection in the x-axis of the graph of $y = \dfrac{1}{x}$.

These are called rectangular hyperbolas.

Exam tip

You must learn the shape of the graphs $y = \dfrac{1}{x}$ and $y = -\dfrac{1}{x}$.

The x-axis is a horizontal asymptote.

The y-axis is a vertical asymptote.

In each case the graph gets closer and closer to the axes without ever touching them.

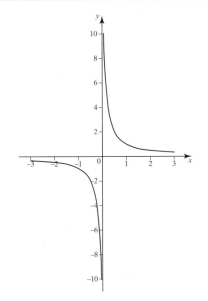

Fig. 1.17
Sketch of graph of $y = \dfrac{1}{x}$.

Using graphs to solve equations

Curve sketching provides a useful method of showing the points of intersection of two (or more) graphs and identifying the number of solutions to an equation.

Example

a) On the same diagram sketch the graphs of the functions $y = x^2 - 4$ and $y = (x + 1)(x - 2)(x - 3)$.

b) How many points of intersection are there?

c) Find the x-coordinates of the points of intersection.

👉 **Continued on the next page**

Exam tips

You will need to recognise and factorise

$x^2 - a^2$ as $(x - a)(x + a)$.

This is known as the difference of two squares.

Answer

a) $y = x^2 - 4$

$x = 0 \Rightarrow y = -4$: the graph passes through $(0, -4)$.

$y = 0 \Rightarrow x^2 - 4 = 0 \Rightarrow (x - 2)(x + 2) = 0$

the graph passes through $(2, 0)$ and $(-2, 0)$.

As x approaches infinity (∞) then y approaches infinity (∞)

As x approaches negative infinity ($-\infty$) then y approaches infinity (∞)

$y = (x + 1)(x - 2)(x - 3)$

$x = 0 \Rightarrow y = 6$: the graph passes through $(0, 6)$

$y = 0 \Rightarrow (x + 1)(x - 2)(x - 3) = 0$

$\Rightarrow x = -1, x = 2$ or $x = 3$

$\Rightarrow$ the graph passes through $(-1, 0)$ $(2, 0)$, $(3, 0)$.

As x approaches infinity (∞) then y approaches infinity (∞).

As x approaches negative infinity ($-\infty$) then y approaches negative infinity ($-\infty$).

Fig. 1.18
Graphs of $y = x^2 - 4$ and $y = (x + 1)(x - 2)(x - 3)$.

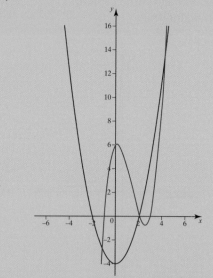

Essential notes

Three points of intersection means there will be three solutions to this equation.

b) Three

c) The x-coordinates of the points of intersection are given as solutions of the equation

$x^2 - 4 = (x + 1)(x - 2)(x - 3)$

$\Rightarrow (x^2 - 4) - (x + 1)(x - 2)(x - 3) = 0$

$\Rightarrow (x - 2)(x + 2) - (x + 1)(x - 2)(x - 3) = 0$

$\Rightarrow (x - 2)((x + 2) - (x + 1)(x - 3)) = 0$

$x = 2$ or $((x + 2) - (x + 1)(x - 3)) = 0$

$x + 2 - (x^2 - 2x - 3) = 0$

$x^2 - 3x - 5 = 0$

$x = \dfrac{3 \pm \sqrt{9 + 20}}{2} = \dfrac{3 \pm \sqrt{29}}{2}$

$x = 2,$

Method notes

Use the formula to solve this quadratic equation to find the second and third roots.

Example

a) On the same diagram sketch the graphs of the functions $y = x^2(x-2)$ and $y = \dfrac{4}{x}$.

b) Use your graphs to decide how many solutions there are to the equation

$$\frac{4}{x} - x^2(x-2) = 0$$

Answer

a) $y = x^2(x-2)$

    $x = 0 \Rightarrow y = 0$: the graph passes through $(0, 0)$

    $y = 0 \Rightarrow x^2(x-2) = 0$

        $\Rightarrow x = 0$ or $x = 2$

        $\Rightarrow$ the graph passes through $(0, 0)$, (a repeated root) $(2, 0)$.

As x approaches infinity (∞) then y approaches infinity (∞).

As x approaches negative infinity $(-\infty)$ then y approaches negative infinity $(-\infty)$.

The shape of the graph of $y = \dfrac{4}{x}$ is similar to the graph of $y = \dfrac{1}{x}$.

It passes through the point $(1, 4)$.

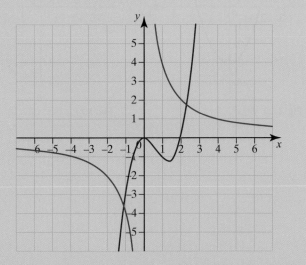

Fig. 1.19
Graphs of $y = x^2(x-2)$ and $y = \dfrac{4}{x}$.

b) From the sketch there are two points of intersection of the two graphs. These points occur where the y-values of each equation are the same i.e.

where $\dfrac{4}{x} = x^2(x-2)$ or where

$$\frac{4}{x} - x^2(x-2) = 0$$

Exam tips

In C1 you will not be expected to solve equations like this other than by graph sketching.

Transformations of graphs

Many graphs of functions can be described in terms of simple **transformations** of the graphs of standard functions. Using transformations often helps us to identify the positions of the key features of a function such as the turning points.

Translation

Figures $1.20 - 1.23$ show the graphs of several quadratic functions. Each figure also shows (in blue) the graph of the 'standard quadratic' function $y = x^2$ which passes through the point $(0, 0)$.

In each figure, look at the transformation of $y = x^2$ in terms of geometrical descriptions.

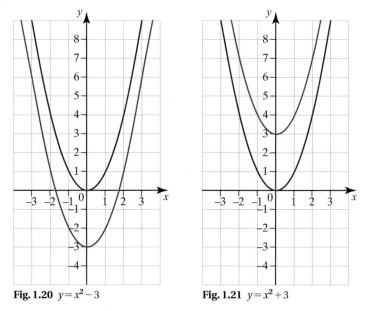

Fig. 1.20 $y = x^2 - 3$　　　　　**Fig. 1.21** $y = x^2 + 3$

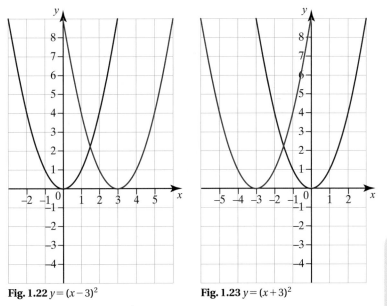

Fig. 1.22 $y = (x-3)^2$ **Fig. 1.23** $y = (x+3)^2$

In Figure 1.20 the graph of $y = x^2$ is translated through -3 units parallel to the y-axis to give $y = x^2 - 3$.

In Figure 1.21 the graph of $y = x^2$ is translated through $+3$ units parallel to the y-axis to give $y = x^2 + 3$.

In Figure 1.22 the graph of $y = x^2$ is translated through $+3$ units parallel to the x-axis to give $y = (x-3)^2$.

In Figure 1.23 the graph of $y = x^2$ is translated through -3 units parallel to the y-axis to give $y = (x+3)^2$.

Method notes

Figure 1.20 shows graph of $y = x^2$ 'moves down' 3 units.

Figure 1.21 shows graph of $y = x^2$ 'moves up' 3 units.

Figure 1.22 shows graph of $y = x^2$ 'moves to the right' 3 units.

Figure 1.23 shows graph of $y = x^2$ 'moves to the left' 3 units.

Method notes

Fig. 1.24 shows $y = 3x^2$ multiplies all the y values of $y = x^2$ by 3. The graph looks like it has been 'pulled inwards'. Fig. 1.25 shows graph being pulled outwards as y values are divided by 3 ($y = \frac{1}{3}x^2$).

Stretch

Suppose now that we multiply the x^2 by a constant. Figures 1.24 − 1.27 show the graphs of several quadratic functions.

In each figure the blue curve is the graph of $y = x^2$

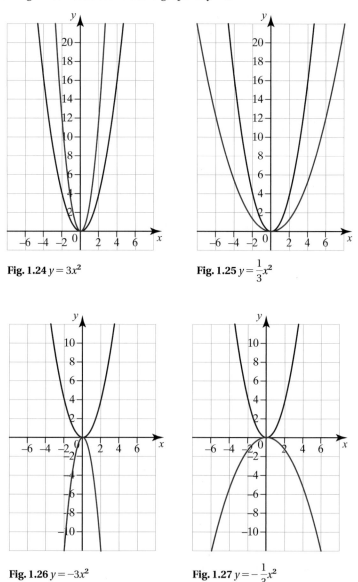

Fig. 1.24 $y = 3x^2$

Fig. 1.25 $y = \dfrac{1}{3}x^2$

Fig. 1.26 $y = -3x^2$

Fig. 1.27 $y = -\dfrac{1}{3}x^2$

Method notes

1. Graph of $y = f(x)$ moves left for $f(x + a)$ pLus moves Left.

2. Graph of $y = f(x)$ moves right for $f(x - a)$ subtRact moves Right.

3. Graph of $y = f(x)$ moves 'up' a units $f(x) + a$.

4. Graph of $y = f(x)$ moves 'down' a units $f(x) - a$.

In Figure 1.24 the graph of $y = x^2$ is stretched by a factor of 3 units parallel to the y-axis to give $y = 3x^2$.

In Figure 1.25 the graph of $y = x^2$ is is stretched by a factor of $\dfrac{1}{3}$ unit parallel to the y-axis to give $y = \dfrac{1}{3}x^2$.

The effect of the negative sign in the functions $y = -3x^2$ and $y = -\dfrac{1}{3}x^2$ is to produce a reflection in the x-axis.

Summary

1. The transformation $f(x + a)$ is a horizontal translation of $-a$ units.
2. The transformation $f(x - a)$ is a horizontal translation of a units.
3. The transformation $f(x) + a$ is a vertical translation of $+a$ units.
4. The transformation $f(x) - a$ is a vertical translation of $-a$ units.

The transformation $af(x)$ is a vertical stretch factor a.

The transformation $f(ax)$ is a horizontal stretch factor $\dfrac{1}{a}$.

Multiple transformations

When two or more transformations are combined, the equation of the transformed curve becomes more complicated.

Example

Figure 1.28 shows the graph of a function $y = f(x)$. The points A, B and C have coordinates (0, 0), (2, 4) and (3, 0) respectively.

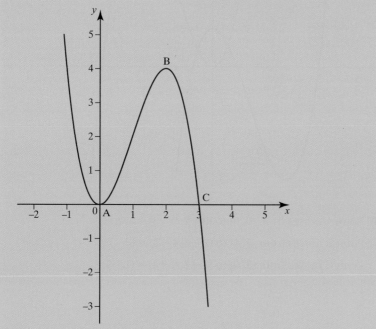

Fig. 1.28
Graph of $y = f(x)$.

Essential notes

A and B are examples of turning points.

A is called a **local minimum** and B is called a **local maximum** of the function.

a) Sketch the graph of the function $y = f(x + 2) - 3$.

b) State the coordinates of the points A, B and C after the transformation.

Continued on the next page

Answer

a) The transformation $f(x + 2)$ is a horizontal translation of -2 units.

The transformation $f(x) - 3$ is a vertical translation of -3 units.

Put these two transformations together and the given curve (blue) is translated horizontally by -2 units followed by a vertical translation of -3 units to give the curve shown in red in Figure 1.29.

Fig. 1.29
Graph of $y = f(x + 2) - 3$ (red curve).

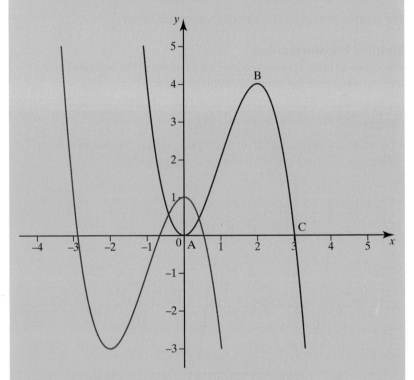

Method notes

Consider each transformation separately. Take the one nearest to x in the equation first. In this case deal with $f(x + 2)$.

The given curve moves 2 units 'to the left'.

Then deal with -3

The given curve moves 'down' 3 units.

Method notes

Apply the transformation $(-2, -3)$ to the coordinates of each point.

b) Point A moves from $(0, 0)$ to $(0 - 2, 0 - 3) = (-2, -3)$

Point B moves from $(2, 4)$ to $(2 - 2, 4 - 3) = (0, 1)$

Point C moves from $(3, 0)$ to $(3 - 2, 0 - 3) = (1, -3)$

Example

On the same diagram, sketch the graphs of the functions with equations $y = -\dfrac{4}{x}$, $x \neq 0$ and $y = 2 - \dfrac{4}{(x - 1)}$, $x \neq 1$.

State clearly the coordinates of any x-intercepts and y-intercepts.

Write down the equations of the asymptotes to the graph of

$$y = 2 - \frac{4}{(x - 1)}, x \neq 1.$$

Exam tip

You must learn the form of the standard algebraic function which gives each transformation.

Answer

The function $y = -\dfrac{4}{x}$ is a vertical stretch of factor 4 $\left(\text{shown by } \dfrac{4}{x}\right)$

followed by a reflection in the x-axis $\left(\text{shown by } -\dfrac{4}{x}\right)$

The function $y = 2 - \dfrac{4}{(x-1)}$ is two translations of the graph of $y = -\dfrac{4}{x}$:

a horizontal translation of $+1$ followed by a vertical translation of $+2$

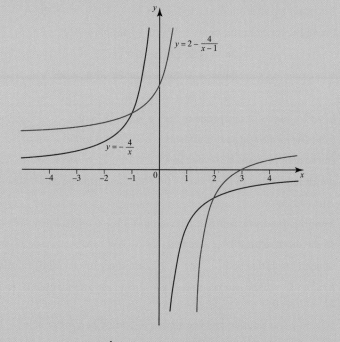

Fig. 1.30
Graphs of $y = -\dfrac{4}{x}$ and $y = 2 - \dfrac{4}{x-1}$.

The graph of $y = 2 - \dfrac{4}{(x-1)}$ cuts the y-axis once where

$x = 0 \Rightarrow y = 2 - \dfrac{4}{(0-1)} = 6$

The y-intercept is $(0, 6)$.

The graph of $y = 2 - \dfrac{4}{(x-1)}$ cuts the x-axis once where

$y = 0 \Rightarrow 0 = 2 - \dfrac{4}{(x-1)} \Rightarrow \dfrac{4}{(x-1)} = 2 \Rightarrow x = 3$

The x-intercept is $(3, 0)$.

$x \to \infty$ the horizontal asymptote of $y = -\dfrac{4}{x}$ is $y = 0$

$x \to \infty$ the horizontal asymptote of $y = 2 - \dfrac{4}{(x-1)}$ is $y = 2$

$y \to \infty$ the vertical asymptote of $y = -\dfrac{4}{x}$ is $x = 0$

$y \to \infty$ the vertical asymptote of the $y = 2 - \dfrac{4}{(x-1)}$ is $x = 1$

Stop and think answers

1. i. Yes if $a = 0$ or $b = 0$

   ii. Yes if $b = 0$ or $a = b = 0$

2. $x^2 + 4x + 4 = (x + 2)^2 + 0$ so whatever value x takes, the answer will be a perfect square (or square number).

3. If $a(a - 4) = b(b - 4)$ then $a^2 - 4a = b^2 - 4b$

   Rearranging as a quadratic in a gives $a^2 - 4a - b^2 + 4b = 0$

   Using the formula: $2a = 4 \pm \sqrt{(16 - 4(-b^2 + 4b))}$

   $$2a = 4 \pm (2b - 4)$$

   $$2a = 4 + 2b - 4 \text{ or } 2a = 4 - 2b + 4$$

   $$a = b \text{ or } a = 4 - b$$

4. i. $2a^2 + 4a = 7$ rearrange to $2a^2 + 4a - 7 = 0$

   from the quadratic solution formula $a = 2$ $b = 4$ and $c = -7$

   so the discriminant $b^2 - 4ac$ is $4^2 - 4(2)(-7) = 16 + 56 = 72$

   which is positive so $b^2 - 4ac > 0$ hence there are two solutions.

   $3b^2 + 1 = 0$

   from the quadratic solution formula $a = 3$ $b = 0$ and $c = 1$

   so the discriminant $b^2 - 4ac$ is $0 - 4(3)(1) = -12$

   which is negative so $b^2 - 4ac < 0$ hence there are no (real) solutions

5. In $x^3 - 1 = 0$ we can see that if $x = 1$ this equation would be true as $1^3 - 1 = 0$

   So $x = 1$ is a solution and therefore $x - 1$ must be a factor of $x^3 - 1 = 0$

   Since the equation is a cubic there are three solutions, not all of which are necessarily real. The following steps investigate this.

   Step 1: Let $x^3 - 1 = (x - 1)(ax^2 + bx + c)$

   Step 2: Expand the brackets

   $$x^3 - 1 = ax^3 + bx^2 + cx - ax^2 - bx - c$$

   Step 3: Compare the coefficients in step 2

   x^3 coefficient gives $a = 1$

   x^2 coefficient gives $b - a = 0$ so $b = 1$

   x coefficient gives $c - b = 0$

   constant gives $-1 = -c$ so $c = 1$

   Hence using these values for a, b and c in step 1:

   $$x^3 - 1 = (x - 1)(x^2 + x + 1)$$

Step 4: Therefore we can write the equation $x^3 - 1 = 0$

as $(x - 1)(x^2 + x + 1) = 0$ so $x - 1 = 0$ or $x^2 + x + 1 = 0$

Step 5: If $x - 1 = 0$ then $x = 1$ is a solution

Step 6: If $x^2 + x + 1 = 0$ the discriminant is $1^2 - 4(1)(1)$ which is negative so there are no real solutions from this.

Hence $x = 1$ is the only (real) solution to $x^3 - 1 = 0$

Essential notes

You should be familiar with these ideas from GCSE Mathematics.

Learn that the equation of a straight line is $y = mx + c$.

Equations of straight lines

Definition

A graph which is a straight line is called a linear graph.

The equation of a straight line is $y = mx + c$ where m is the slope or gradient of the line and c is the y-intercept, i.e. the line passes through the point $(0, c)$.

> **Example**
> a) Write down the equation of the straight line with gradient 3 and y-intercept -5.
>
> b) Write down the gradient and y-intercept of the lines with equations
>
>   i) $y = 2x - 7$
>
>   ii) $y = -4x + 8$
>
> **Answer**
> a) The straight line with gradient 3 and y-intercept -5 has $m = 3$ and $c = -5$
>
> The equation of the straight line is therefore $y = 3x - 5$
>
> b) i) $y = 2x - 7 \Rightarrow m = 2$ and $c = -7$
>
>   ii) $y = -4x + 8 \Rightarrow m = -4$ and $c = 8$

Figure 2.1 shows a linear graph through two points (x_1, y_1) and (x_2, y_2) and the gradient of the line is m.

Fig. 2.1
Linear graph.

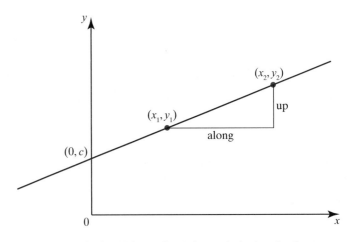

The gradient m is calculated from the right angled triangle shown.

$$m = \frac{\text{up}}{\text{along}} = \frac{\text{change in } y}{\text{change in } x} = \frac{y_2 - y_1}{x_2 - x_1}$$

Essential notes

Another word for gradient is 'slope'.

m is the shorthand notation used for the gradient of a straight line.

Example

Three points A, B and C have coordinates A(−1, −2), B(4, 6) and C(2, 8) respectively.

a) Find the gradient of the straight line through A and B.

b) Find the gradient of the straight line through B and C.

Answer

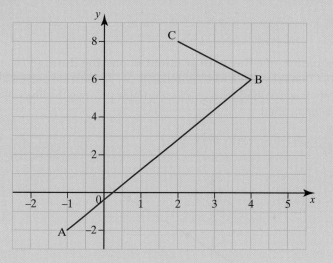

Fig. 2.2
Points A, B and C.

Method notes

It is useful to draw a diagram showing the points and then joining the points to show the lines.

a) gradient of AB $= \dfrac{6 - (-2)}{4 - (-1)} = \dfrac{8}{5} = 1.6$

b) gradient of BC $= \dfrac{8 - 6}{2 - 4} = \dfrac{2}{-2} = -1$

The signs of the gradients are different. The gradient of AB is positive whereas the gradient of BC is negative. This is important.

A line which slopes **upwards** from left to right has a **positive** gradient.

A line which slopes **downwards** from left to right has a **negative** gradient.

Exam tip

Learn the difference between lines with a positive gradient and those with a negative gradient.

Remember that in the form $ax + by + c = 0$

a is the x-coefficient

b is the y-coefficient

c is the constant term.

Different ways of writing the equation of a straight line

Consider the line with equation $y = -4x + 8$

Rearranging this algebraically: add $4x$ to both sides

$y + 4x = 8$

Subtracting 8 from both sides: $y + 4x - 8 = 0$ which is an alternative way of writing the equation.

This means that any equation of a straight line can be written in rearranged form as $ax + by + c = 0$

Method notes

Substitute for the values of m and c.

Multiply by 2

Rearrange the formula.

Example

A straight line has gradient 0.5 and y-intercept -1.5

a) Write the equation in the form $y = mx + c$.

b) Write the equation in the form $ax + by + c = 0$

Answer

a) $y = 0.5x - 1.5 = \dfrac{1}{2}x - \dfrac{3}{2}$

b) $2y = x - 3 \Rightarrow 2y - x + 3 = 0$

The equation of a line with gradient m passing through (x_1, y_1)

Suppose that the point A with coordinates (x_1, y_1) is a fixed point and the point P with coordinates (x, y) is a variable point on the line with a gradient m. Then

$$\frac{y - y_1}{x - x_1} = m \Rightarrow y - y_1 = m(x - x_1)$$

Fig. 2.3
Line with a gradient m.

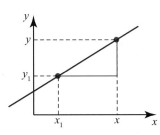

This gives yet another way of writing the equation of a straight line. It should be used for finding the equation if you know the gradient and a point on the line.

Example

Find the equation of the straight line with gradient 2 which passes through the point $(-1, 6)$.

Answer

Substituting for $x_1 = -1$ and $y_1 = 6$

$$y - 6 = 2(x - (-1)) \Rightarrow y - 6 = 2x + 2$$

$$\Rightarrow y = 2x + 8 \quad \text{or} \quad y - 2x - 8 = 0$$

Essential notes

Points are **collinear** if they lie on the same straight line.

You must learn that there are three ways of finding the equation of a straight line:

Use $y = mx + c$ if given the gradient and intercept

Use $y - y_{11} = m(x - x_{11})$ if given the gradient and one point on the line

Use $\dfrac{y - y_1}{y_2 - y_1} = \dfrac{x - x_1}{x_2 - x_1}$ if given two points on the line.

The equation of a line passing through (x_1, y_1) and (x_2, y_2)

Let the points $A(x_1, y_1)$ and $B(x_2, y_2)$ be fixed points and the point P (x, y) be a variable point on the line through A and B. Since A, B and P are collinear:

Then gradient AP = gradient AB

$$\frac{y - y_1}{x - x_1} = \frac{y_2 - y_1}{x_2 - x_1} \Rightarrow \frac{y - y_1}{y_2 - y_1} = \frac{x - x_1}{x_2 - x_1}$$

This gives yet another way of writing the equation of a straight line. It should be used for finding the equation if you know two points on the line.

Example
Find the equation of the straight line through the following pairs of points (2, 0) and (7, 1).

Answer
Substituting for $x_1 = 2$, $y_1 = 0$, $x_2 = 7$ and $y_2 = 1$

$$\frac{y - 0}{1 - 0} = \frac{x - 2}{7 - 2} \Rightarrow y = \frac{x - 2}{5}$$

$$\Rightarrow y = \frac{1}{5}x - \frac{2}{5} \quad \text{or} \quad 5y - x + 2 = 0$$

Stop and think 1

1. *What would be the equation of the line parallel to the x-axis passing through the point (0, b)?*

2. *What would be the equation of the line parallel to the y-axis passing through the point (a, 0)?*

3. *What would be the equation of the line passing through the origin and the point (a, a)?*

4. *What would be the equation of the line passing through the origin and the point (b, b)?*

5. *Why are the answers to questions 3 and 4 the same?*

Parallel lines

Definition
When two lines are **parallel** they have **equal gradients**.

Exam tip
Learn this definition.

Example
Four points have coordinates A(3, 5), B(8, 3), C(6, −2), D(−4, 2).

Show that AB is parallel to CD.

Answer

gradient of AB $= \dfrac{3 - 5}{8 - 3} = \dfrac{-2}{5} = -0.4$

gradient of CD $= \dfrac{2 - (-2)}{(-4) - 6} = \dfrac{4}{(-10)} = -0.4$

Since the gradients are the same, the two lines AB and CD are parallel.

Method notes
Remember that the gradient
$$= \frac{y_2 - y_1}{x_2 - x_1}$$

Perpendicular lines

Two lines are perpendicular to each other when they intersect at right angles i.e. the angle between them is 90°. We can find the relationship between their gradients.

As an example, Figure 2.4a shows a line through the origin with gradient 2. Figure 2.4b shows a rotation of the line through an angle of 90°. The triangle OAB rotates to the triangle OCD.

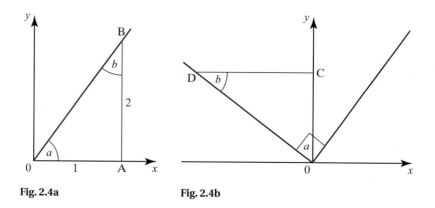

Fig. 2.4a   **Fig. 2.4b**

The angles in triangle OAB are equal in size to the angles in triangle OCD e.g. angle COD = angle AOB = a

Essential notes

Rotation does not change the size of the triangle.

Two triangles whose angles are equal in size and whose sides have the same length. are said to be **congruent**.

The line OD is a rotation of the line OB so that the length of OD equals the length of OB.

The two triangles OAB and OCD are the same size.

Thus CD = 2 and OC = 1

Gradient of OB = $\dfrac{2}{1} = 2$

Gradient of OD = $\dfrac{\text{OC}}{\text{CD}} = \dfrac{1}{-2} = -\dfrac{1}{2}$

For the two perpendicular lines gradient of OB × gradient of OD = −1

This result can be generalised. Figure 2.5 shows two perpendicular lines PQ and RS.

Fig. 2.5
Perpendicular lines.

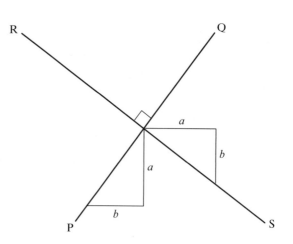

Gradient of PQ $= \dfrac{a}{b}$ We call this m (1)

Gradient of RS $= -\dfrac{b}{a}$ so we can see that this is $-\dfrac{1}{m}$ (2)

If two straight lines are perpendicular, (1) $\times$ (2) shows the product of their gradients is -1

Exam tip

You must learn this result.

Example

Find an equation of the line:

a) which is perpendicular to the line $y = 2x + 5$, passing through the point $(-1, 4)$.

b) which is perpendicular to the line $3x + y - 6 = 0$, passing through the point $(6, 1)$.

Answer

a) The gradient of the line $y = 2x + 5$ is 2

The gradient of the perpendicular line is $-\dfrac{1}{2}$

The equation of the perpendicular line through $(-1, 4)$ is

$$\dfrac{y - 4}{x + 1} = -\dfrac{1}{2} \Rightarrow y - 4 = -\dfrac{1}{2}(x + 1)$$

$$\Rightarrow y = -\dfrac{1}{2}x + 3\dfrac{1}{2} \quad \text{or} \quad 2y + x - 7 = 0$$

b) $3x + y - 6 = 0 \Rightarrow y = -3x + 6$ when rearranged.

The gradient of the line $y = -3x + 6$ is -3

The gradient of the perpendicular line is $\dfrac{1}{3}$

The equation of the perpendicular line through $(6, 1)$ is

$$\dfrac{y - 1}{x - 6} = \dfrac{1}{3} \Rightarrow y - 1 = \dfrac{1}{3}(x - 6)$$

$$\Rightarrow y = \dfrac{1}{3}x - 1 \quad \text{or} \quad 3y - x + 3 = 0$$

Method notes

Always plot the points and label them correctly before joining any points.

Fig. 2.6
Rectangle ABCD.

Example

A quadrilateral has vertices A(0, 10), B(8, −2), C(2, −6) and D(−6, 6). By working out gradients, show that ABCD is a rectangle.

Answer

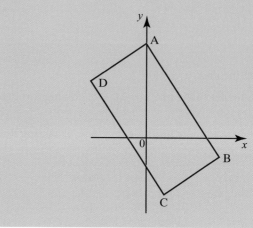

$$\text{Gradient of AB} = \frac{-2-10}{8-0} = -\frac{12}{8} = -\frac{3}{2}$$

$$\text{Gradient of BC} = \frac{-6-(-2)}{2-8} = \frac{-4}{-6} = \frac{2}{3}$$

$$\text{Gradient of CD} = \frac{6-(-6)}{(-6)-2} = \frac{12}{-8} = -\frac{3}{2}$$

$$\text{Gradient of DA} = \frac{10-6}{0-(-6)} = \frac{4}{6} = \frac{2}{3}$$

Since their gradients are the same, AB is parallel to CD.

Since their gradients are the same, BC is parallel to DA.

Since gradient AB × gradient BC is $-\frac{3}{2} \times \frac{2}{3} = -1$

and gradient CD × gradient DA is $-\frac{3}{2} \times \frac{2}{3} = -1$

AB is perpendicular to BC and DA.

CD is perpendicular to BC and DA.

So the opposite sides of ABCD are parallel and adjacent sides of ABCD are perpendicular, so the quadrilateral is a rectangle.

Stop and think 2

In the rectangle ABCD in the example, are the diagonals AC and BD perpendicular to each other?

Give reasons for your answer.

Stop and think answers

1 1 Line is parallel to the x-axis (it is a horizontal line) so gradient $m = 0$

It passes through $(0, b)$ so the y-intercept $c = b$

Form of equation to use is $y = mx + c$

so $y = b$ is the equation

2 Line is parallel to y-axis (it is a vertical line) so all points on this line will be the same distance from the y-axis i.e. they will have the same x-coordinates. As line passes through $(a, 0)$ all x-coordinates of points on this line will be a so $x = a$ is the equation.

It is also true that the gradient of this line is infinity.

3 Line passes through 2 points (a, a) and $(0, 0)$.

Gradient of line m is therefore $\dfrac{a - 0}{a - 0} = 1$

Passing through $(0, 0)$ shows the intercept $c = 0$

So equation is $y = x$ or rearranged it is $y - x = 0$

4 Same method as part 3 above using (b, b) gives $m = 1$ and $c = 0$

So equation is $y = x$ or $y - x = 0$

5 The answers to 3 and 4 describe the same line since the 3 points $(0, 0)$, (a, a) and (b, b) are collinear. If you take any two of the three points given and work out the gradient of the line joining them it will always be $= 1$

2 Gradient $AC = \dfrac{-6 - 10}{2 - 0} = -8$

Gradient $BD = \dfrac{6 - -2}{-6 - 8} = \dfrac{8}{-14}$

If you multiply these two answers together you do not get –1 so the diagonals are not perpendicular.

Mathematics is often described as the 'science of patterns'. When solving problems in mathematics, particular patterns of numbers that are obtained from the data give rise to rules or formulas which form part of the solution to the problem. The patterns of numbers often form sequences or series. This chapter explores the properties of particular examples: arithmetic sequences and series.

Sequences: notation and language

Definitions

A sequence is an ordered list of numbers which follows a set rule.

Each number in a sequence is called a term.

The set rule must show a relationship between one term and the next term.

The terms of any sequence are often written using the following notation:

$u_1, u_2, u_3, u_4, u_5, \ldots u_n, \ldots$ which is a shorthand way of describing the sequence where:

u_1 is the first term,

u_2 is the second term,

u_3 is the third term,

u_n is the nth term etc.

You can use any suffix letter to describe the general term of the sequence, but the most commonly used letters are k or n. The final term is l.

Example

For each of the following sequences

i) find the next three terms

ii) find a rule of generating the next number from the ones before it.

a) 4, 8, 12, 16, 20, …

b) 11, 8, 5, 2, −1, …

c) 1, 2, 4, 8, 16, …

d) 243, 81, 27, 9, 3, …

e) 1, 4, 9, 16, 25, …

f) 1, 1, 2, 3, 5, 8, …

Answer

a) 4, 8, 12, 16, 20, …

Step 1: Try to find a relationship between any two consecutive terms which is true throughout the sequence.

Step 2: In this sequence the difference between consecutive terms is equal to 4

Step 3: this means

$u_1 = 4$

$u_2 = u_1 + 4 = 4 + 4 = 8$

$u_3 = u_2 + 4 = 8 + 4 = 12$

$u_4 = u_3 + 4 = 12 + 4 = 16$

$u_5 = u_4 + 4 = 16 + 4 = 20$, etc.

In general $u_{k+1} = u_k + 4$.

Step 4: The next three terms are $u_6 \; u_7 \; u_8$

Use $u_{k+1} = u_k + 4$.

$u_6 = u_5 + 4 = 20 + 4 = 24$

$u_7 = 24 + 4 = 28$

$u_8 = 28 + 4 = 32$

b) 11, 8, 5, 2, −1, …

The difference between each term is equal to −3

In general $u_{k+1} = u_k - 3$

So the next three terms are

$u_6 = u_5 - 3 = -4$

$u_7 = u_6 - 3 = -7$

$u_8 = u_7 - 3 = -10$

c) 1, 2, 4, 8, 16, …

In this sequence each term is double the previous term, so the next three terms are 32, 64, 128

so in general $u_{k+1} = 2 \times u_k$.

d) 243, 81, 27, 9, 3, …

In this sequence each term is one third of the previous term, so the next three terms are $1, \dfrac{1}{3}, \dfrac{1}{9}$

so in general $u_{k+1} = \dfrac{1}{3} \times u_k$

e) 1, 4, 9, 16, 25, …

In this sequence the difference between each term forms a familiar sequence 3, 5, 7, 9 etc. (i.e. the sequence of odd numbers) so the next three terms are 36 (add 11), 49 (add 13), 64 (add 15)

so in general $u_1 = 1$ and $u_{k+1} = u_k + (2k + 1)$

f) 1, 1, 2, 3, 5, 8, …

In this sequence the sum of the first two terms gives the third term; the sum of the second and third terms gives the fourth term; etc. so the next three terms are

13 (5 + 8), 21 (8 + 13) and 34 (13 + 21)

so in general $u_1 = 1$, $u_2 = 1$, and $u_{k+1} = u_k + u_{k-1}$

Method notes

a) Because all these terms are multiples of 4 you may also have spotted a simple rule for the nth term: $u_n = 4n$

Method notes

b) You may have spotted a simple rule for the nth term as $u_n = 14 - 3n$

Method notes

c) You may have spotted a simple rule for the nth term as $u_n = 2^{n-1}$

Method notes

d) A formula for this sequence may not be so obvious! The rule for the nth term as $u_n = 243 \times \dfrac{1}{3}^{\,n-1}$

Method notes

e) These are all square numbers and a simple rule for the nth term is $u_n = n^2$

Method notes

f) This sequence is called a **Fibonacci sequence**. There is a rule for the nth term but it is beyond the scope of C1.

Essential notes

Arithmetic sequences are part of the C1 course.

You will **not** be expected to find the general rule for the nth term of sequences such as c) d) e) and f).

In sequences a) and b) of the last example there is a constant difference between consecutive terms. Such sequences are called **arithmetic sequences**. Their properties are explored in this chapter.

Sequences c) d) e) and f) in the example are not arithmetic sequences.

Example

The nth term of a sequence is given by $u_n = 4n + 3$

a) Write down the first three terms of the sequence.

b) Write down the 24th term of the sequence.

Answer

a) $u_1 = 4 \times 1 + 3 = 7$ Substitute $n = 1$

    $u_2 = 4 \times 2 + 3 = 11$ Substitute $n = 2$

    $u_3 = 4 \times 3 + 3 = 15$ Substitute $n = 3$

b) $u_{24} = 4 \times 24 + 3 = 99$ Substitute $n = 24$

This is an example of an arithmetic sequence.

The iterative process

Writing one term in terms of the term before it, and specifying the first term is called an **inductive definition**. The formula which relates consecutive terms is called a **recurrence relationship** or a **recurrence formula** or an **iterative formula**.

Example

For the following recurrence relation, write down the values of u_2, u_3 and u_4 given that

$$u_1 = 2, \ u_{k+1} = \frac{1}{2}u_k + 6$$

Essential notes

A value for u_1 must be given to get the sequence started.

Answer

Step 1: Write down the value of u_1

    $u_1 = 2$ starts off the sequence.

Step 2: Use $u_{k+1} = \frac{1}{2}u_k + 6$ to generate the sequence

Step 3: $u_2 = \frac{1}{2}u_1 + 6$

    $\Rightarrow u_2 = \frac{1}{2}(2) + 6 = 7$ (using value in step 1)

    $u_2 = 7 \Rightarrow u_3 = \frac{1}{2}u_2 + 6 = \frac{7}{2} + 7 = \frac{21}{2}$

Step 4: $u_3 = \frac{21}{2} \Rightarrow u_4 = \frac{1}{2}u_3 + 6 = \frac{21}{4} + 6 = \frac{45}{4}$

Arithmetic sequences

Given the general term of a sequence as $u_n = 4n + 3$ the first few terms of the sequence would be 7, 11, 15, 19, 23, …We can see that this sequence has a common difference of 4 between successive terms. This means it is an example of an arithmetic sequence.

Definition

An arithmetic sequence is a sequence of numbers with a common difference between successive terms.

In general, an arithmetic sequence with a first term a and common difference d would have these first four terms

$a, a + d, a + d + d, a + d + d + d, …$

This simplifies to $a, a + d, a + 2d, a + 3d, …$

So we can write this as

$u_1 = a$

$u_2 = a + d$

$u_3 = a + 2d$

$u_4 = a + 3d$ which gives

$u_n = a + (n - 1)d$

This gives the general term of an arithmetic sequence using the first term and the common difference.

Essential notes

An arithmetic sequence is also called an **arithmetic progression** or A.P.

Essential notes

Remember that the number multiplying d is one less than the 'term number'.

For u_4 multiply d by 3 etc.

Exam tip

This formula for u_n is given in your formulae booklet.

Example

Write down the 15th term and the nth term of the sequence 8, 13, 18, 23, …

Answer

The first term $a = 8$ and the common difference $d = 5$

The 15th term $u_{15} = a + 14d = 8 + 14 \times 5 = 78$

The nth term $u_n = a + (n - 1)d = 8 + 5(n - 1) = 5n + 3$

Example

The 5th term of a sequence is 17 and the 12th term is 52. Find the common difference, first term and nth term of the sequence.

Answer

The 5th term $u_5 = a + 4d = 17$ and the 12th term $u_{12} = a + 11d = 52$

Solving the two simultaneous equations:

$7d = 52 - 17$

$\quad = 35$

$\quad d = 5, a = -3$

So $u_n = a + (n - 1)d = -3 + (n - 1)5 = 5n - 8$

Method notes

Solving simultaneous equations was covered in Chapter 1

Example
Find the number of terms in the arithmetic sequence 26, 30, ... 66, 70, 74

Answer
The first term $a = 26$ and the common difference $d = 4$

Let there be n terms in the sequence so the final term is

$$u_n = a + (n-1)d = 26 + 4(n-1) = 74$$

$$\Rightarrow 4(n-1) = 74 - 26 = 48$$

$$\Rightarrow 4n = 52 \Rightarrow n = 13$$

which means there are 13 terms in the sequence.

Series

Definition
A series is formed when the terms of a sequence are added together. 1, 3, 5, 7, 9, 11 is a sequence of six terms. $1 + 3 + 7 + 9 + 11$ is the corresponding series for this sequence.

Example
Due to the change in government financing, a solar power company needs to complete the installations of solar panels in a four-week period. One hundred people start work on the first Monday and are paid a flat rate of £480 for a week. Each Monday the company will need to employ an **extra** 40 employees at the same rate of pay. Find:

a) how many people are employed at the end of the four weeks

b) how much has the company paid out in wages during the project.

Answer
a) At the start of week 1 there were 100 employees.

At the start of week 2 there were $100 + 40$ employees.

At the start of week 3 there were $140 + 40$ employees.

The number of employees each week will form an arithmetic sequence where $u_1 = 100$, $u_2 = 140$, $u_3 = 180$, $u_4 = 220$

At the end of the four weeks the company is employing 220 people.

b) The total wages paid out by the company is the sum of the weekly workforce multiplied by £480

which means $480 (u_1 + u_2 + u_3 + u_4)$

$$= £480 \times 100 + £480 \times 140 + £480 \times 180 + £480 \times 220$$

$$= £48\ 000 + £67\ 200 + £86\ 400 + £105\ 600$$

$$= £307\ 200$$

During the project the company has paid £307 200 in wages.

The sum of the salaries is an example of an **arithmetic series**. Before investigating the properties of arithmetic series there is further notation to understand.

Partial sums

> **Example**
> For the arithmetic sequence 3, 6, 9, 12, 15, 18, 21 form a series by adding the following terms:
>
> a) $u_1 + u_2$ b) $u_1 + u_2 + u_3$
>
> c) $u_1 + u_2 + u_3 + u_4$ d) $u_1 + u_2 + u_3 + u_4 + u_5$
>
> **Answer**
> a) $u_1 + u_2 = 3 + 6 = 9$
>
> b) $u_1 + u_2 + u_3 = 3 + 6 + 9 = 18$
>
> c) $u_1 + u_2 + u_3 + u_4 = 3 + 6 + 9 + 12 = 30$
>
> d) $u_1 + u_2 + u_3 + u_4 + u_5 = 3 + 6 + 9 + 12 + 15 = 45$

Each of the sums in the example is called a **partial sum** of the series 3, 6, 9, 12, 15, 18, 21.

> **Stop and think**
>
> *Which of the terms in the sequence 3, 3.5, 4, 4.5, 5, 5.5 have an integer value?*
>
> *Does the partial sum $u_2 + u_3 + u_4$ have an integer answer?*

Sigma notation

$S_1 = u_1$ is called the first partial sum.

$S_2 = u_1 + u_2$ is called the second partial sum.

$S_3 = u_1 + u_2 + u_3$ is called the third partial sum.

$S_n = u_1 + u_2 + u_3 + \ldots + u_n$ is called the nth partial sum.

The symbol Σ is often used to denote partial sums. It provides a convenient short hand notation.

For example,

a) $\displaystyle\sum_{k=1}^{10} u_k$ means the sum of the first 10 terms of a sequence u_1, u_2, u_3 up to u_{10} because the notation means k must take all integer values from 1 to 10

so $\displaystyle\sum_{k=1}^{10} u_k = u_1 + u_2 + u_3 + u_5 \ldots + u_{10}$

Essential notes

The symbol Σ is the Greek letter **sigma**.

The label on Σ must link with the letter used to define the sequence.

Essential notes

$\sum_{r=1}^{6} r^2$ uses the letter r as a **variable** which is an integer.

Method notes

Substitute for $r=1, 2, 3, 4, 5, 6$ to find the terms in the series.

b) $\sum_{r=1}^{6} r^2 = 1^2 + 2^2 + 3^2 + 4^2 + 5^2 + 6^2 = 91$ (r must take the integer values 1, 2, 3, 4, 5, 6). It also means that the general term is r^2 so $\sum_{r=1}^{6} r^2 =$ the sum of the first six square numbers.

Example

Calculate $\sum_{r=1}^{6} (3 + 2r)$.

Answer

$$\sum_{r=1}^{6} (3+2r) = 5+7+9+11+13+15 = 60$$

Method notes

In this example the values of k start at 4.

It is important to look at the lower and upper limits carefully. These values show with which terms the series starts and finishes.

Example

Calculate $\sum_{k=4}^{7} (k^2 - 2)$.

Answer

Step 1: Write out the kth term: $u_k = k^2 - 2$

Step 2: State the k values: $k = 4, 5, 6, 7$

Step 3: Work out $u_4 = 4^2 - 2 = 14$ then u_5, u_6 and u_7

Step 4: Add these terms together to give the answer.

$$\sum_{k=4}^{7} (k^2 - 2) = (4^2 - 2) + (5^2 - 2) + (6^2 - 2) + (7^2 - 2)$$

$$= 14 + 23 + 34 + 47 = 118$$

Example

A sequence is defined by $u_1 = 2$, $u_{n+1} = 5 - 2u_n$ for $n \geq 1$

a) Find u_2 and u_3

b) Calculate $\sum_{k=1}^{5} u_k$ and show that the sum is a multiple of 3

Answer

a) $u_2 = 5 - 2u_1 = 1$

$u_3 = 5 - 2u_2 = 3$

b) $\sum_{k=1}^{5} u_k = u_1 + u_2 + u_3 + u_4 + u_5 = 2 + 1 + 3 + (-1) + 7 = 12$

12 is a multiple of 3 and so $\sum_{k=1}^{5} u_k$ is a multiple of 3

Arithmetic series

Definition

An arithmetic series is defined as the sum of the terms of an arithmetic sequence.

In an earlier section we found that if a is the first term and d is the common difference the general form of an arithmetic sequence is given by:

$$u_1 = a, \ u_2 = a + d, \ u_3 = a + 2d, \ u_4 = a + 3d, \ ...$$

$$u_n = a + (n - 1)d$$

The general form of an arithmetic series of n terms (since a series is a sum of terms) is therefore given by

$$S_n = a + (a + d) + (a + 2d) + ... + (a + (n - 1)d)$$

Essential notes

An arithmetic sequence is a sequence with a common difference.

Essential notes

Notice that the nth term has $(n - 1)$ multiples of d.

Example

An arithmetic series is formed from an arithmetic sequence with first term 3 and common difference 7

a) Find the 15th term of the series.

b) Find the first term in the series which is larger than 1000

Answer

a) The arithmetic sequence has $a = 3$ and $d = 7$ so the general term is
$u_n = 3 + 7(n - 1)$ for $n \geq 1 \Rightarrow u_1 = 3, \ u_2 = 10$

The arithmetic series is therefore $3 + 10 + 17 + 24 + ...$

The 15th term of the series $u_{15} = 3 + 7(15 - 1) = 101$

b) We need $u_n > 1000 \Rightarrow 3 + 7(n - 1) > 1000 \Rightarrow 3 + 7n - 7 > 1000$
$\Rightarrow 7n > 1004$

The smallest integer n for which $7n > 1004$ is $n = 144$

The first term in the series which is larger than 1000 is
$u_{144} = 3 + 7(144 - 1) = 1004$

Method notes

We are only interested in integer solutions of $7n > 1004$ because n is an integer.

The method of finding the sum of an arithmetic series is often attributed to the famous German mathematician and scientist Carl Friedrich Gauss (1777–1855). At the age of seven, Gauss was asked to sum the integers from 1 to 100. His teacher was amazed when Gauss spotted that the sum was equal to 50 pairs of numbers with each pair summing to 101. So the sum is 5050.

The following example demonstrates what Gauss must have observed.

Example

Find the sum of the first 100 **even** numbers.

Answer

Let $\qquad S = 2 + 4 + 6 + \ldots + 196 + 196 + 200$

In reverse $\quad S = 200 + 198 + 196 + \ldots + 6 + 4 + 2$

Adding corresponding terms in the sums above

$$2S = 202 + 202 + 202 + \ldots + 202 + 202 + 202$$

There are 100 terms all equal to 202

So $\qquad 2S = 100 \times 202$

$$S = 50 \times 202 = 10\,100, \text{ giving the answer.}$$

The approach in this example gives a general method of adding any arithmetic series.

Consider the arithmetic series

$$S_n = a + (a + d) + (a + 2d) + \ldots + (a + (n - 2)d) + (a + (n - 1)d)$$

In reverse

$$S_n = (a + (n - 1)d) + (a + (n - 2)d) + (a + (n - 3)d) + \ldots + (a + d) + a$$

Add the two sums: term 1 to term 1 then term 2 to term 2 and so on.

$$2S_n = [a + a + (n - 1)d] + [a + d + a + (n - 2)d] + \ldots$$

But there are n of these terms all equal to $[2a + (n - 1)d]$

$$\Rightarrow 2S_n = n \times [2a + (n - 1)d]$$

$$\boldsymbol{S_n = \frac{n}{2} \times [2a + (n - 1)d]}$$

Knowing the first term, the common difference and the number of terms, the sum of an arithmetic series can be found using this formula.

Exam tip

You need to learn the proof of the sum of an arithmetic series and remember the final formula.

Example

Find the sum of the first 20 terms of the following series.

a) $3 + 8 + 13 + 18 + 23 + \ldots$

b) $6 + 2 - 2 - 6 - 10 - \ldots$

Answer

a) For the series $3 + 8 + 13 + 18 + 23 + \ldots$, $a = 3$, $d = 5$

$$S_{20} = \frac{20}{2} \times [2 \times 3 + (20 - 1) \times 5] = 1010$$

b) For the series $6 + 2 - 2 - 6 - 10 - \ldots$, $a = 6$, $d = -4$

$$S_{20} = \frac{20}{2} \times [2 \times 6 + (20 - 1) \times (-4)] = -640$$

Method notes

Use the general formula:

$S_n = \frac{n}{2} \times [2a + (n - 1)d]$

Example

The sixth term of an arithmetic series is 21 and the sum of the first ten terms is 185.

Find the common difference and the first term of the sequence forming the series.

Answer

Sixth term means $n = 6$ so $u_6 = a + 5d = 21$

S_{10} means the sum of the first 10 terms and
$S_{10} = 5 \times (2a + 9d) = 185 \Rightarrow 2a + 9d = 37$

Solving the simultaneous equations $a + 5d = 21$ and $2a + 9d = 37$ gives $d = 5$ and $a = -4$

Method notes

Use general term
$u_n = a + (n - 1)d$

Use sum of n terms
$S_n = \dfrac{n}{2} \times [2a + (n - 1)d]$

Solution of simultaneous equations was covered in Chapter 1

There is an alternative formula for the sum of the first n terms of an arithmetic series.

$$2a + (n - 1)d = a + [a + (n - 1)d]$$
$$= u_1 + u_n$$
$$= \text{first term} + \text{last term}$$

$$S_n = \frac{n}{2} \times (\text{first term} + \text{last term})$$

Example

a) Find the sum of the first 1000 multiples of 3

b) Find the sum of the arithmetic series $-12 + \ldots + 4$ which has 11 terms.

Answer

a) The series formed from the multiples of 3 is $3 + 6 + 9 + 12 + 15 + 18 + \ldots + 3000$

first term $= 3$, last term $= 3000$ with $n = 1000$

$$S_{1000} = \frac{1000}{2} \times (3 + 3000) = 1\,501\,500$$

b) first term $= -12$, last term $= 4$ with $n = 11$

$$S_{11} = \frac{11}{2} \times (-12 + 4) = -44$$

Method notes

Which of the two formulas you use for S_n depends on the information given in the question.

If you are given the number of terms and the first and last term use

$$S_n = \frac{n}{2} \times (\text{first} + \text{last term})$$

If you are given the number of terms, the first term and the common difference use

$$S_n = \frac{n}{2} \times [2a + (n - 1)d]$$

Stop and think answer

An integer is a whole number so the 1st, 3rd and 5th terms have integer values.

The partial sum required is $3.5 + 4 + 4.5 = 12$ which is an integer so the answer is yes.

Differentiation is an important part of a branch of mathematics called **calculus**. Differentiation describes the **rate of change** of a quantity.

Speed is a familiar concept. We are all used to travelling, whether on foot, by car, by bus or by train, and we often talk about the speed at which we are travelling. Suppose that a person travels by car from Wells (in Somerset) to Plymouth (in Devon), a distance of 110 miles in 2 hours. Then the average speed for the journey is 55 m.p.h. The average speed for a journey is calculated as the ratio

$$\text{average speed} = \frac{\text{distance travelled}}{\text{time taken}}$$

However, the average speed is unlikely to be the actual speed as shown by a car's speedometer. This will very from 0 m.p.h. when the car is stopped at traffic lights or in traffic queues, to 70 m.p.h. when the car is cruising along the M5 motorway. The reading of the speedometer is the speed at an instant and is called the **instantaneous** speed. Average speed and instantaneous speed are familiar examples of rates of change.

Rate of change is an important idea in many areas of life such as science, engineering, economics and business. For example, the growth and decline in the economy can be described in terms of rates of change. Not all rates of change are calculated with respect to time. We are often interested in the rate of change with respect to other quantities such as distance. For example, when finding the optimum size of a cylindrical container we would investigate the rate of change of volume with respect to radius.

The concept and process of differentiation are best understood through investigating the slope or gradient of a curve at any point along the curve.

Gradient of a curve and the gradient function

The meaning of the gradient to a curve
The **slope** or **gradient** of a straight line can be positive, negative or zero.

Essential notes

The words slope and gradient mean the same thing.

Fig. 4.1
Linear functions.

Example
Find the gradients of the following linear functions.

(a)　　　　　　　　(b)

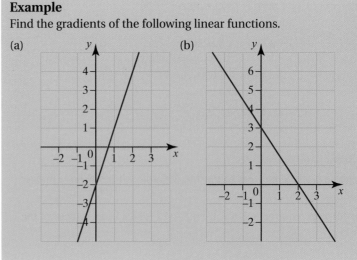

(c) $y = 1.7x + 2.4$ (d) $y = -2.7x + 4.2$

Answer

(a) gradient $= \dfrac{6}{2} = 3$ A positive gradient

(b) gradient $= -\dfrac{3}{2} = -1.5$ A negative gradient

(c) gradient $= 1.7$

(d) gradient $= -2.7$

Method notes

The gradient of a straight line graph is the ratio

$$\frac{\text{difference in } y\text{-coordinates}}{\text{difference in } x\text{-coordinates}}$$

In a linear equation of the form $y = mx + c$ the gradient is the number that multiplies x.

Example

What can you say about the gradients at the points A, B, C, D and E of the graph in Figure 4.2 below?

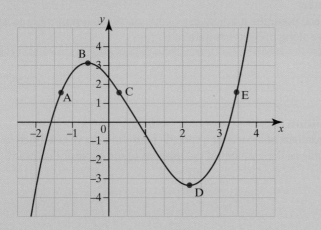

Fig. 4.2
Cubic function.

Answer

In this case, the gradient of the graph changes as you move along it. It is more difficult to give a value for the gradient at each point. What we can say is that:

at A the gradient is positive

at B the gradient is zero; this point is called a local maximum

at C the gradient is negative

at D the gradient is zero; this point is called a local minimum

at E the gradient is positive.

Essential notes

Points B and D are examples of **stationary points** or **turning points** where the gradient = 0

We use the word 'local' because these points are not actual maximum or minimum values. For example, the graph ends at a point which is higher that point B.

☞ **Continued on the next four pages**

Fig. 4.3

Example
What can you say about the gradients at the points A, B, C, D and E of the graph in Figure 4.3?

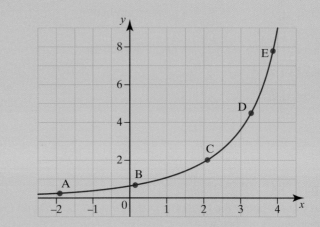

Answer
At each point the gradient is positive and increases in value as we move along the curve from A to E.

Definition
The **gradient** of a curve at a specific point on a curve is defined as being the same as the gradient of the **tangent** to the curve at that point.

Fig. 4.4
Tangent to a curve at P.

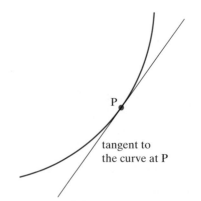

The tangent to a curve is a straight line which touches the curve at a point and never cuts the curve. An approximate method of drawing the tangent would be to try to lay a straight edge (such as a ruler) along the curve so that the ruler is a tangent. Then calculate the gradient of the ruler! This method leads to very approximate values.

A more accurate method is to 'zoom in' on the point so that the curve appears to straighten. Eventually the curve becomes a straight line and then you can find the gradient of the straight line.

Method notes

You can do this zooming using a graphic calculator or computer software.

Fig. 4.5
Zooming in on a curve.

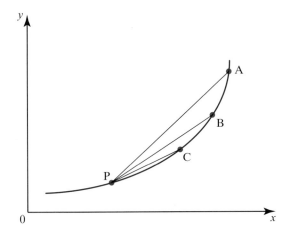

The process is represented in Figure 4.5. The gradient of the chord PA would be an approximate value of the gradient of the tangent. As the point which starts at A moves to B and then C getting closer to P then the gradients of the chords PB, PC, etc. become closer to the gradient of the tangent at the point P. This provides a method for finding the gradient of the curve at P.

Example
Figure 4.6 shows the graph of the equation $y = x^2$. The point P has coordinates (1, 1). Points A, B and C have coordinates (1.5, 2.25), (1.25, 1.5625) and (1.1, 1.21) respectively.

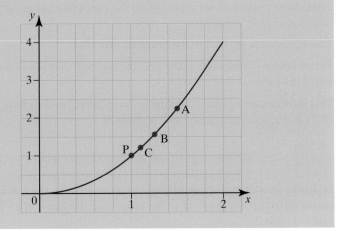

Fig. 4.6
Graph of the equation $y = x^2$.

☞ **Continued on the next two pages**

Method notes

Gradient formula is

$$\frac{\text{difference in } y-\text{coordinates}}{\text{difference in } x-\text{coordinates}}$$

Calculate the gradients of the lines joining P to A, P to B and from P to C.

Answer

$$\text{gradient of PA} = \frac{2.25 - 1}{1.5 - 1} = \frac{1.25}{0.5} = 2.5$$

$$\text{gradient of PB} = \frac{1.5625 - 1}{1.25 - 1} = \frac{0.5625}{0.25} = 2.25$$

$$\text{gradient of PC} = \frac{1.21 - 1}{1.1 - 1} = \frac{0.21}{0.1} = 2.1$$

In the above example the gradients of the chords PA, PB and PC are decreasing. The values of the gradients look as if they are approaching the value 2.

If we continue the process by taking the points D (1.01, 1.0201), E (1.001, 1.002001) and F (1.0001, 1.00020001) we are zooming in closer to the point P (1, 1) and the curve becomes a straight line i.e. the tangent. The gradients from each of these points to P are:

- gradient of PD $= \dfrac{1.0201 - 1}{1.01 - 1} = \dfrac{0.0201}{0.01} = 2.01$

- gradient of PE $= \dfrac{1.002001 - 1}{1.001 - 1} = \dfrac{0.002001}{0.001} = 2.001$

- gradient of PF $= \dfrac{1.00020001 - 1}{1.0001 - 1} = \dfrac{0.00020001}{0.0001} = 2.0001$

The gradient is very close to 2. If we continued the process, taking points even closer to point P, we would get to the exact value of the gradient, which is actually equal to 2.

This introduces the idea of the **gradient function**.

Each of the x-coordinates of the points close to P that were chosen above are of the form $1 + h$.

This provides an algebraic approach. The gradient of the chord from (1, 1) to the point $(1 + h, (1 + h)^2)$ is given by

Essential notes

This process is called finding the gradient from first principles.

$$\frac{(1 + h)^2 - 1}{(1 + h) - 1} = \frac{1 + 2h + h^2 - 1}{h} = \frac{2h + h^2}{h} = 2 + h$$

As h gets smaller and smaller then $2 + h$ gets closer to the value 2. So we can deduce that the gradient of the tangent to the graph of $y = x^2$ at the point (1, 1) is 2.

Example

Find the gradient of the tangent to the graph of $y = x^2$ at the points (2, 4) and (3, 9).

Answer

The gradient of the chord from (2, 4) to the point $(2 + h, (2 + h)^2)$

is given by: $\dfrac{\text{difference in } y-\text{coordinates}}{\text{difference in } x-\text{coordinates}}$ giving

$$\frac{(2 + h)^2 - 4}{(2 + h) - 2} = \frac{4 + 4h + h^2 - 4}{h} = \frac{4h + h^2}{h} = 4 + h$$

As h gets smaller and smaller then $4 + h$ gets closer to the value 4. So we can deduce that the gradient of the tangent to the graph of $y = x^2$ at the point (2, 4) is 4.

The gradient of the chord from (3, 9) to the point $(3 + h, (3 + h)^2)$ is given by

$$\frac{(3 + h)^2 - 9}{(3 + h) - 3} = \frac{9 + 6h + h^2 - 9}{h} = \frac{6h + h^2}{h} = 6 + h$$

As h gets smaller and smaller then $6 + h$ gets closer to the value 6. So we can deduce that the gradient of the tangent to the graph of $y = x^2$ at the point (3, 9) is 6

Method notes

$(2 + h)^2 = 2^2 + 4h + h^2$

The following table shows the results for the gradient of the tangents to the graph of $y = x^2$ at the three points (1, 1), (2, 4) and (3, 9).

point	gradient of the tangent
(1, 1)	2
(2, 4)	4
(3, 9)	6

The gradient function

At each point P of the graph of a function $y = f(x)$ the gradient of the tangent at P will depend on the x-coordinate of P. So the gradient defines a new function called the **gradient function** or the derived function. The form of this new function depends on the given function f(x).

The gradient function or the derived function is denoted by f'(x).

A formal definition of the gradient function is given by:

f '(x) is the limit of $\dfrac{f(x + h) - f(x)}{h}$ as $h \to 0$

Method notes

For each point on the curve $y = x^2$, the gradient is twice the x-coordinate of the point.

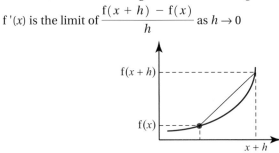

Fig. 4.7
The gradient function.

We now apply this definition to a few simple functions.

Finding the gradient function for f(x) = x^2

$$\frac{f(x + h) - f(x)}{h} = \frac{(x + h)^2 - x^2}{h} = \frac{x^2 + 2xh + h^2 - x^2}{h}$$

$$= \frac{2xh + h^2}{h} = \frac{h(2x + h)}{h} = 2x + h$$

As h tends to zero, $2x + h$ tends to $2x$.

$f(x) = x^2 \Rightarrow f'(x) = 2x$

Finding the gradient function for f(x) = x^3

$$\frac{f(x + h) - f(x)}{h} = \frac{(x + h)^3 - x^3}{h} = \frac{x^3 + 3x^2h + 3xh^2 + h^3 - x^3}{h}$$

$$= \frac{3x^2h + 3xh^2 + h^3}{h} = \frac{h(3x^2 + 3xh + h^2)}{h} = 3x^2 + 3xh + h^2$$

As h tends to zero, $3x^2 + 3xh + h^2$ tends to $3x^2$

$f(x) = x^3 \Rightarrow f'(x) = 3x^2$

Finding the gradient function for f(x) = x^4

$$\frac{f(x + h) - f(x)}{h}$$

$$= \frac{(x + h)^4 - x^4}{h} = \frac{x^4 + 4x^3h + 6x^2h^2 + 4xh^3 + h^4 - x^4}{h}$$

$$= \frac{4x^3h + 6x^2h^2 + 4xh^3 + h^4}{h} = \frac{h(4x^3 + 6x^2h + 4xh^2 + h^3)}{h}$$

$$= 4x^3 + 6x^2h + 4xh^2 + h^3$$

As h tends to zero, $4x^3 + 6x^2h + 4xh^2 + h^3$ tends to $4x^3$

$f(x) = x^4 \Rightarrow f'(x) = 4x^3$

Finding the gradient function for f(x) = $x^2 - 7x + 10$

$$\frac{f(x + h) - f(x)}{h} = \frac{(x + h)^2 - 7(x + h) + 10 - (x^2 - 7x + 10)}{h}$$

$$= \frac{x^2 + 2xh + h^2 - 7x - 7h + 10 - x^2 + 7x - 10}{h} = \frac{2xh + h^2 - 7h}{h}$$

$$= 2x - 7 + h$$

As h tends to zero, $2x - 7 + h$ tends to $2x - 7$

$f(x) = x^2 - 7x + 10 \Rightarrow f'(x) = 2x - 7$

Rules of differentiation

The gradient function for powers of x

The results from the previous four examples gave us several gradient functions:

f(x)	x^2	x^3	x^4	$x^2 - 7x + 10$
f'(x)	$2x$	$3x^2$	$4x^3$	$2x - 7$

These results suggest some important patterns which can be used to find gradient functions. $f(x) = x^3 \Rightarrow f'(x) = 3x^{3-1} = 3x^2$ which leads to the conclusion:

$f(x) = x^n \Rightarrow f'(x) = nx^{n-1}$

e.g. $f(x) = x^5 \Rightarrow f'(x) = 5x^{5-1} = 5x^4$ hence:

$f(x) = kx^n \Rightarrow f'(x) = knx^{n-1}$, **where k is a constant.**

e.g. $f(x) = 7x^5 \Rightarrow f'(x) = 7 \times 5x^{5-1} = 35x^4$. The constant multiplier 7 remains and rule of differentiation is used on x^5

The gradient function of a sum (or difference) of functions is the sum (or difference) of the gradient functions of the individual functions.

e.g. $f(x) = x^5 - 4x^3 \Rightarrow f'(x) = 5x^{5-1} - 4 \times 3x^{3-1} = 5x^4 - 12x^2$

Example

Find the gradient function for each of the following f(x):

(a) $f(x) = x^2 + x$

(b) $f(x) = x^4 + 3x - 7.5$

(c) $f(x) = 3x(x^2 - 4)$

(d) $f(x) = \dfrac{x^2 - 3x + 4}{x}$

(e) $f(x) = x^{\frac{1}{2}}$

(f) $f(x) = \dfrac{1}{x^2}$

Answer

(a) $f(x) = x^2 + x \Rightarrow f'(x) = 2x + 1$

(b) $f(x) = x^4 + 3x - 7.5 \Rightarrow f'(x) = 4x^3 + 3$

(c) $f(x) = 3x(x^2 - 4) \Rightarrow f(x) = 3x^3 - 12x$

$\Rightarrow f'(x) = 9x^2 - 12$

(d) $f(x) = \dfrac{x^2 - 3x + 4}{x} \Rightarrow f(x) = \dfrac{x^2}{x} - \dfrac{3x}{x} + \dfrac{4}{x} = x - 3 + \dfrac{4}{x}$

$\Rightarrow f(x) = x - 3 + 4x^{-1}$

$\Rightarrow f'(x) = 1 + 4 \times (-1x^{-1-1}) = 1 - 4x^{-2} = 1 - \frac{4}{x^2}$

(e) $f(x) = x^{\frac{1}{2}} \Rightarrow f'(x) = \frac{1}{2}x^{\frac{1}{2}-1} = \frac{1}{2}x^{-\frac{1}{2}}$

(f) $f(x) = \frac{1}{x^2} \Rightarrow f(x) = x^{-2}$

$\Rightarrow f'(x) = -2x^{-2-1} = -2x^{-3} = -\dfrac{2}{x^3}$

Gradient function notation and first order derivative

Until now we have used function notation:

f(x) for the original function $\Rightarrow$ f'(x) for the gradient function.

The gradient function is also represented by $\dfrac{dy}{dx}$, which is not a fraction, but a single quantity.

$$y = f(x) \Rightarrow \frac{dy}{dx} = f'(x)$$

The notation $\dfrac{dy}{dx}$ comes from the idea that the gradient of a chord is formed from taking two 'small steps or increments' dx in the x-direction and dy in the y-direction as shown in Figure 4.8 below and then allowing dx → 0

Fig. 4.8
Gradient of a chord.

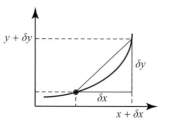

Essential notes

dx is called 'delta x'.

dy is called 'delta y'.

The gradient of the curve is then the limit of $\dfrac{dy}{dx}$ as $x \rightarrow 0$.

Essential notes

The rules for finding $\dfrac{dy}{dx}$ are the same as the rules for finding f'(x) as $\dfrac{dy}{dx} = f'(x)$.

For powers of x, $y = x^n$, the gradient function is $\dfrac{dy}{dx} = nx^{n-1}$

The process of finding the gradient function or first order derivative is called differentiation with respect to x.

Example

Find $\dfrac{dy}{dx}$ for the following functions.

(a) $y = x^7$ (b) $y = 13x + 5$

(c) $y = 4x^2 - 9x + 3$ (d) $y = x^{\frac{2}{3}} - 5x^2$

Answer

(a) $y = x^7$ $\Rightarrow \dfrac{dy}{dx} = 7x^6$

(b) $y = 13x + 5$ $\Rightarrow \dfrac{dy}{dx} = 13$

(c) $y = 4x^2 - 9x + 3$ $\Rightarrow \dfrac{dy}{dx} = 8x - 9$

(d) $y = x^{\frac{2}{3}} - 5x^2$ $\Rightarrow \dfrac{dy}{dx} = \dfrac{2}{3}x^{-\frac{1}{3}} - 10x$

Essential notes

In (b) the constant 5 rewritten ready for using the rule of differentiation is $5x^0$. Therefore using the rule the gradient function of 5 is

$0x^{0-1} = 0$

In (c) similarly the constant 3 becomes 0 when differentiated.

Example
Find the points on the curve $y = x^3 - 5x^2 + 5x - 3$ where the gradient is 2

Answer
$\dfrac{dy}{dx}$ gives the gradient so:

$$y = x^3 - 5x^2 + 5x - 3 \quad \Rightarrow \quad \dfrac{dy}{dx} = 3x^2 - 10x + 5$$

Where the gradient is 2, $\dfrac{dy}{dx} = 3x^2 - 10x + 5 = 2$

So $3x^2 - 10x + 3 = 0$

$\Rightarrow (3x - 1)(x - 3) = 0$

$\Rightarrow \qquad\qquad x = \frac{1}{3}$ or $x = 3$

$x = \dfrac{1}{3} \quad \Rightarrow y = \left(\dfrac{1}{3}\right)^3 - 5\left(\dfrac{1}{3}\right)^2 + 5\left(\dfrac{1}{3}\right) - 3 = -\dfrac{50}{27}$

$x = 3 \quad \Rightarrow y = (3)^3 - 5(3)^2 + 5(3) - 3 = -6$

The points on the curve of $y = x^3 - 5x^2 + 5x - 3$ where the gradient is 2 are $\left(\dfrac{1}{3}, -\dfrac{50}{27}\right)$ and $(3, -6)$.

Notation
In this chapter we have introduced several expressions for the process of differentiation. This may seem a bit confusing!

You do not need to panic!

The following words essentially mean the same thing:

what we say	what we do
find the derivative of x^2	differentiate to give $2x$
find the gradient of the tangent to $y = x^2$ at any point	differentiate to give $2x$
find the slope of the tangent to $y = x^2$ at any point	differentiate to give $2x$
differentiate x^2	differentiate to give $2x$

The second order derivative

Given a function $y = f(x)$ the gradient function $\dfrac{dy}{dx} = f'(x)$ gives a new function of x. This function can also be differentiated.

$$\dfrac{dy}{dx} = \dfrac{d^2y}{dx^2} = f''(x) \text{ is called the } \textbf{second order derivative.}$$

Example

Find $\dfrac{d^2y}{dx^2}$ for $y = 5x^4 + 3x^2 - \dfrac{4}{x^2}$

Answer

$$y = 5x^4 + 3x^2 - \frac{4}{x^2} = 5x^4 + 3x^2 - 4x^{-2} \Rightarrow \frac{dy}{dx} = 20x^3 + 6x + 8x^{-3}$$

$$\frac{d^2y}{dx^2} = 60x^2 + 6 - 24x^{-4} = 60x^2 + 6 - \frac{24}{x^4}$$

Fig. 4.9
Graph of $y = (x+1)(x-2)(x-3)$.

Stop and think 1

Figure 4.9 shows the graph of the function $y = (x+1)(x-2)(x-3)$.

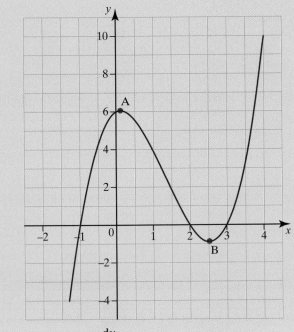

What can you say about $\dfrac{dy}{dx}$ at points A and B?

Use differentiation to find the x-coordinates of the points A and B.

Exam tip

Displacement is distance travelled in a direction. You are expected to know that the symbol for displacement is s, for velocity is v, for acceleration is a and for time it is t. You also need to know the following connections between them.

The rate of change of displacement (s) with time (t)

$$= \frac{ds}{dt} = v.$$

The rate of change of velocity (v) with time (t)

$$\frac{dv}{dt} = a.$$

Applications of differentiation

Rates of change

At the start of this chapter we introduced the idea of differentiation in terms of the rate of change of quantities, such as distance, with respect to time. Rate of change of distance is instantaneous velocity. In such situations the letters used for the variables are unlikely to be x and y.

For example, a stone thrown vertically in the air can be described in terms of the height of the stone above the ground as a function of time,

$h = f(t)$. The velocity would then be $v = \dfrac{dh}{dt} = f'(t)$.

The rules for differentiating powers of t are the same as differentiating powers of x.

For powers of t, $h = t^n$, the gradient function is $\dfrac{dh}{dt} = nt^{n-1}$

Example

Rachel throws a ball straight up in the air. The height, s metres, of the ball above the ground is given by the formula $s = 1.4 + 12t - 5t^2$ where t is the time in seconds.

(a) Find the velocity of the ball after one second.

(b) Find the greatest height reached by the ball.

(c) Find the time when the ball hits the ground and the velocity of the ball when it hits the ground.

Answer

(a) $s = 1.4 + 12t - 5t^2 \Rightarrow$ velocity of ball $v = \dfrac{ds}{dt} = 12 - 10t$

when $t = 1$, $v = \dfrac{ds}{dt} = 12 - 10 = 2$

The speed of the ball after one second is 2 m s^{-1}

(b) The greatest height of the ball occurs when $v = 0$.

$\Rightarrow v = \dfrac{ds}{dt} = 12 - 10t = 0 \Rightarrow t = 1.2$

The greatest height reached by the ball occurs after 1.2 seconds.

(c) The ball hits the ground when $s = 0$.

$\Rightarrow s = 1.4 + 12t - 5t^2 = 0$

$\Rightarrow t = \dfrac{-12 \pm \sqrt{12^2 - 4 \times (-5) \times 14}}{-10} = \dfrac{-12 \pm \sqrt{172}}{-10}$

$\Rightarrow t = -0.11$ or $t = 2.51$

The ball hits the ground after 2.51 seconds.

The velocity of the ball when it hits the ground is

$v = 12 - 10 \times 2.51 = -13.1$

The ball hits the ground with speed 13.1 m s^{-1}

Method notes

1.4 is a constant and so differentiates to 0

After 1 second is when $t = 1$

Greatest height reached is when the ball is instantaneously at rest so $v = 0$

Ball hits ground when its overall displacement $s = 0$

Use the quadratic formula $\dfrac{-b \pm \sqrt{b^2 - 4ac}}{2a}$ with $a = -5$, $b = 12$ and $c = 1.4$

Reject the negative solution $t = -0.11$ as times are always positive.

Example

The displacement of a particle is given by $s = t^4 + 3t - 7.5$ where s is measured in metres and t is the time in seconds. Find the velocity and acceleration after 2 seconds.

Answer

$s = t^4 + 3t - 7.5 \Rightarrow$ velocity of particle $v = \dfrac{ds}{dt} = 4t^3 + 3$

$\Rightarrow$ acceleration of particle $v = \dfrac{dv}{dt} = 12t^2$

Method notes

Differentiate $s(t)$ to give a formula for $v(t)$.

Differentiate $v(t)$ to give a formula for $a(t)$.

Continued on the next page

After 2 seconds, velocity $= 4(2)^3 + 3 = 35$ m s^{-1}

and acceleration $= 12(2)^2 = 48$ m s^{-2}

Essential notes

$\frac{dy}{dx} = f'(x)$ when $x = a$ is written $f'(a)$.

Finding the equation of a tangent to a curve

The gradient of the tangent to a curve $y = f(x)$ at a point $(a, f(a))$ is found by evaluating $\frac{dy}{dx} = f'(x)$ when $x = a$ then using $y - y_1 = m(x - x_1)$.

The equation of a straight line that forms the tangent is then given by $y - f(a) = f'(a)(x - a)$.

Method notes

First find the gradient function.

Then find the gradient at the point $(-1, 3)$.

Finally use the equation of the straight line of gradient m through the given point (x_1, y_1)

$y - y_1 = m(x - x_1)$.

Example

Find the equation of the tangent to the curve $y = 1 - 4x + 2x^3$ at the point $(-1, 3)$.

Answer

$y = 1 - 4x + 2x^3 \Rightarrow \frac{dy}{dx} = -4 + 6x^2$

At the point $(-1, 3)$ $\frac{dy}{dx} = -4 + 6(-1)^2 = 2$

The gradient of the tangent at the point $(-1, 3)$ is 2

The equation of the tangent through $(-1, 3)$ is

$y - 3 = 2(x - (-1)) = 2(x + 1) \Rightarrow y = 2x + 5$

Essential notes

Perpendicular means 'at right angles to' 'or 90° to'.

Finding the equation of a normal to a curve

The **normal** is closely related to the tangent to a curve. The normal at a point on a curve is a straight line which passes through the point and is **perpendicular** to the tangent at the point.

If two lines are perpendicular to each other then the product of their gradients is -1. The product of the gradient of a normal and its tangent is -1

If the gradient of the tangent to a curve is m then the gradient of the normal at the same point is $-\frac{1}{m}$

Method notes

The normal is a straight line so use $y - y_1 = m(x - x_1)$ to find its equation.

Example

Find the equation of the normal to the curve $y = x + \frac{1}{x}$ at the point $(2, 2.5)$.

Answer

$y = x + \frac{1}{x} = x + x^{-1} \Rightarrow \frac{dy}{dx} = 1 - x^{-2}$

At the point $(2, 2.5)$ $\frac{dy}{dx} = 1 - 2^{-2} = \frac{3}{4}$

The gradient of the tangent at the point (2, 2.5) is $\dfrac{3}{4}$

The gradient of the normal at the point (2, 2.5) is $\dfrac{1}{\frac{3}{4}} = -\dfrac{4}{3}$

The equation of the normal through (2, 2.5) is

$$y - 2.5 = -\frac{4}{3}(x - 2)$$

$$\Rightarrow \quad y = -\frac{4}{3}x + \frac{8}{3} + \frac{5}{2} = -\frac{4}{3}x + \frac{31}{6}$$

$$\Rightarrow \quad 6y = 8x + 31$$

Exam tip

Either form of writing the equation of the normal is acceptable.

Stop and think answers

At A and B $\dfrac{\mathrm{d}y}{\mathrm{d}x} = 0$ as the tangent drawn at these points would be horizontal.

This means the slope of the tangent is 0

We are given $\qquad y = (x + 1)(x - 2)(x - 3)$

Multiplying out gives $y = (x + 1)(x^2 - 5x + 6)$

$$= x^3 + x^2 - 5x^2 - 5x + 6x + 6$$

So $\qquad\qquad\qquad y = x^3 - 4x^2 + x + 6$

Differentiate $\dfrac{\mathrm{d}y}{\mathrm{d}x} = 3x^2 - 8x + 1$ \hfill (1)

At A and B $\qquad\qquad \dfrac{\mathrm{d}y}{\mathrm{d}x} = 0$ \hfill (2)

So equating (1) and (2) $\qquad 3x^2 - 8x + 1 = 0$

Using the formula to solve with $\quad a = 3 \quad b = -8 \quad c = 1$

Gives $\quad x = \dfrac{8 + \sqrt{52}}{6}$ and $\dfrac{8 - \sqrt{52}}{6}$

Which simplifies to $x = \dfrac{8 + 2\sqrt{13}}{6}$ and $\dfrac{8 - 2\sqrt{13}}{6}$ using rules of surds.

So the x-coordinates in exact form are

$\dfrac{4 + \sqrt{13}}{3}$ and $\dfrac{4 - \sqrt{13}}{3}$

Introduction

Integration is another branch of calculus. Although it is often introduced as the reverse process of differentiation, the first steps in developing the techniques of integration were taken by mathematicians in ancient Greece – long before differentiation was developed by mathematicians in the 17th century. Archimedes, around 225 BC, made one of the most significant contributions when finding an approximation to the area of a circle. This is an early example of integration and led to approximate values of π. Among other 'integrations' by Archimedes were the volume and surface area of a sphere, the volume and area of a cone and the surface area of an ellipse.

An application of the process of integration is in finding the area between the line on a graph and the x-axis. We will use the simple straight line graph of $y = mx + c$ between $x = a$ and $x = b$ as illustrated in Figure 5.1 to introduce the important notation for integration which is used throughout this chapter.

Fig. 5.1
Finding an area.

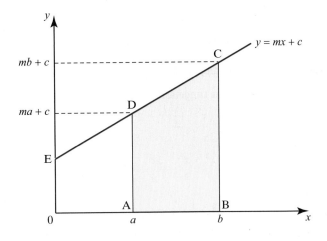

Essential notes

$OA = a$

$OB = b$

Method notes

Point E is on the line $y = mx + c$

This means that given the x-coordinate of a point on the line we can find the y-coordinate at that point using this line equation.

At E $x = 0$ so $y = 0 + c$

Therefore OE is of length c.

Area of a trapezium $= \frac{1}{2}$ (sum of lengths of parallel sides) × perpendicular distance between them

The area of the trapezium 0ADE is $\frac{1}{2}(c + (ma + c))a$

The area of the trapezium 0BCE is $\frac{1}{2}(c + (mb + c))b$

Subtracting the two expressions, the area of the trapezium ABCD is

$$\frac{1}{2}(c + (mb + c))b - \frac{1}{2}(c + (ma + c))a$$

$$= \frac{1}{2}(2cb + mb^2) - \frac{1}{2}(2ca + ma^2)$$

$$= \left(\frac{1}{2}mb^2 + cb\right) - \left(\frac{1}{2}ma^2 + ca\right)$$

Notice that the expression in each bracket is of the same form. If we write

$A(x) = \frac{1}{2}mx^2 + cx$, then $A(b) = \frac{1}{2}mb^2 + cb$ and $A(a) = \frac{1}{2}ma^2 + ca$

$\Rightarrow$ Area of ABCD $= A(b) - A(a)$

The function A(x) is called the area function, and for the function
f(x) = $mx + c$, A(x) = $\frac{1}{2}mx^2 + cx$

Definitions

The area function is also called the **integral function**.

A(x) = $\frac{1}{2}mx^2 + cx$ is called the **integral** of the function f(x) = $mx + c$.

This method of finding the area of the trapezium enclosed by the graph
of y = f(x) = $mx + c$ between the x-axis and the lines $x = a$ and $x = b$
using the integral function process is called **integration**.

This area is written as $[A(x)]_a^b = A(b) - A(a)$

This is called **definite integration** since it has finite value. We know the
values of a and b.

Integration as the inverse process of differentiation

Although developed separately over approximately two thousand years,
the processes of differentiation and integration are closely linked.

Step 1: Write out the area function

A(x) = $\dfrac{1}{2}mx^2 + cx$

Step 2: Differentiate to obtain the result:

$\dfrac{dA}{dx} = mx + c$ which is also equal to f(x).

Step 3: Rewrite f(x) = $\dfrac{dA}{dx}$

The integral of a function f(x) is denoted by a special symbol, $\int$.

Integration is the inverse process of differentiation.

This means that the integral of the equation in step 2 will give the
equation in step 1.

Step 4: Using mathematical notation this means we write

$\int f(x)\,dx = A(x)$

Step 5: We stated at the beginning of this section that

f(x) = $mx + c$ and A(x) = $\frac{1}{2}mx^2 + cx$

Step 6: Use the results from step 4 to give

$\int mx + c\,dx = \frac{1}{2}mx^2 + cx$

Step 7: We can conclude from step 6 that

$\int mx\,dx = \frac{1}{2}mx^2$ and $\int c\,dx = cx$

Essential notes

This is called the
Fundamental Theorem
of Calculus.

Notation

The elongated S $\int$ symbol
means 'integrate'. The
expression to be integrated
is f(x). dx tells you which
variable is to be integrated:
here it is x.

Rules of integration

The rules of differentiation are used to develop the rules of integration.

For example, if $y = x^3$ then $\dfrac{dy}{dx} = 3x^2$ (1)

But if $y = x^3 + 5$ which means $y = x^3 + 5x^0$

$$\frac{dy}{dx} = 3x^2 + 5(0x^{0-1}) = 3x^2 + 0$$

therefore if $y = x^3 + 5$ then $\dfrac{dy}{dx} = 3x^2$ (2)

This is the same answer as in (1) above!

Constant of integration

From (1) and (2) above we can conclude that any constant added to $y = x^3$ and then differentiated will give the same answer as if we just differentiated $y = x^3$.

Therefore, if $y = x^3 + c \implies \dfrac{dy}{dx} = 3x^2$ and c can be any constant.

Definition: Integration is the process of finding y when given the function $\dfrac{dy}{dx}$.

We know that integration is the inverse process of differentiation.

This means that integrating $\dfrac{dy}{dx} = 3x^2$ will give $y = x^3 + c$ and   (3)

differentiating $y = x^3 + c$ will give $\dfrac{dy}{dx} = 3x^2$ (4)

Using mathematical notation for (3) above we write

$$\int \frac{dy}{dx}\, dx = \int 3x^2 dx \text{ will give } y = x^3 + c$$

Using mathematical notation for (4) above we write

$$\frac{d}{dx}(x^3 + c) = \frac{dy}{dx} = 3x^2$$

Similarly if we started with

$\dfrac{dy}{dx} = 4x^3$ using integration $y = x^4 + c$ (5)

$\dfrac{dy}{dx} = 5x^4$ using integration $y = x^5 + c$ (6)

We can generalise this to a rule of integration.

$$\frac{dy}{dx} = x^n \text{ then } y = \frac{1}{n+1}x^{n+1} + c$$

Example

Find y for each of the following.

a) $\dfrac{dy}{dx} = x^6$ b) $\dfrac{dy}{dx} = x^{-4}$ c) $\dfrac{dy}{dx} = x^{\frac{2}{3}}$

d) $\dfrac{dy}{dx} = 5x^6$ e) $\dfrac{dy}{dx} = 0.8x^3$ f) $\dfrac{dy}{dx} = 6x^{-\frac{1}{2}}$

Answer

a) $\dfrac{dy}{dx} = x^6$

$y = \dfrac{x^7}{7} + c$

b) $\dfrac{dy}{dx} = x^{-4}$

$y = \dfrac{x^{-3}}{-3} + c = -\dfrac{x^{-3}}{3} + c$

c) $\dfrac{dy}{dx} = x^{\frac{2}{3}}$

$y = \dfrac{x^{\frac{5}{3}}}{\left(\frac{5}{3}\right)} + c = \dfrac{3}{5}x^{\frac{5}{3}} + c$

d) $\dfrac{dy}{dx} = 5x^6 = 5 \times x^6$

$y = 5 \times \dfrac{x^7}{7} + c = \dfrac{5x^7}{7} + c$

e) $\dfrac{dy}{dx} = 0.8x^3 = 0.8 \times x^3$

$y = 0.8 \times \dfrac{x^4}{4} + c = 0.2x^4 + c$

f) $\dfrac{dy}{dx} = 6x^{-\frac{1}{2}}$

$y = 6 \times \dfrac{x^{\frac{1}{2}}}{\frac{1}{2}} + c = 12x^{\frac{1}{2}} + c$

Method notes

Use the rule of integration

a) $\dfrac{dy}{dx} = x^n$ with $n = 6$

$y = \dfrac{x^{6+1}}{6+1} + c$

b) with $n = -4$, $n + 1 = -3$
remember to add 1

c) $n = \dfrac{2}{3}$, $n + 1 = \dfrac{2}{3} + 1 = \dfrac{5}{3}$

Remember that $\dfrac{1}{\frac{5}{3}} = \dfrac{3}{5}$

For d), e) and f)

If the function of x is multiplied by a constant then you only need to consider the power of x when integrating.

$\dfrac{dy}{dx} = kx^n \rightarrow y = \dfrac{k}{n+1}x^{n+1} + c$

$n \neq -1$ and k is a constant.

Stop and think 1

Why does $\int c\,dx = cx + m$ where c and m are constants?

Different notations for integration

In the previous section we used the notation:

given $\dfrac{dy}{dx}$ then find the formula for y.

Several different types of notation are used in calculus. The basic rule for integrating powers of x applies in each case.

Essential notes

From the chapter on differentiation, if $f(x) = x^3$ then $f'(x) = 3x^2$.

Example

This example uses the f'(x) notation.

Find f(x) for each of the following.

a) $f'(x) = 3x$ b) $f'(x) = 4x^{-\frac{3}{2}}$

Answer

a) $f'(x) = 3x \implies f(x) = 3 \times \dfrac{x^2}{2} + c = \dfrac{3x^2}{2} + c$

b) $f'(x) = 4x^{-\frac{3}{2}} \implies f(x) = 4 \times \dfrac{x^{-\frac{1}{2}}}{\left(-\frac{1}{2}\right)} + c = -8x^{-\frac{1}{2}} + c$

Example

This example uses the integral sign notation.

Find the expressions for the following integrals.

a) $\int 6x^2 \, dx$ b) $\int 3\sqrt{x} \, dx$

Answer

a) $\int 6x^2 \, dx = 6 \times \dfrac{x^3}{3} + c = 2x^3 + c$

b) $\int 3\sqrt{x} \, dx = \int 3x^{\frac{1}{2}} \, dx = 3 \times \dfrac{x^{\frac{3}{2}}}{\frac{3}{2}} + c = 2x^{\frac{3}{2}} + c$

Method notes

$\sqrt{x} \equiv x^{\frac{1}{2}}$

Be careful with the arithmetic of fractions!

Integrating algebraic expressions

Example

These examples include sums of functions.

Find y for each of the following

a) $\dfrac{dy}{dx} = x^2 - 3x^{\frac{1}{2}} - 12x^{-2}$

b) $\dfrac{dy}{dx} = -2x^{-3} + 3x^3 + x^{\frac{1}{3}}$

Answer

a) $\dfrac{dy}{dx} = x^2 - 3x^{\frac{1}{2}} - 12x^{-2}$

$y = \dfrac{x^3}{3} - 3\dfrac{x^{\frac{3}{2}}}{\frac{3}{2}} - 12\dfrac{x^{-1}}{-1} + c = \dfrac{x^3}{3} - 2x^{\frac{3}{2}} + 12x^{-1} + c$

b) $\dfrac{dy}{dx} = -2x^{-3} + 3x^3 + x^{\frac{1}{3}}$

$y = -2\dfrac{x^{-2}}{-2} + \dfrac{3x^4}{4} + \dfrac{x^{\frac{4}{3}}}{\frac{4}{3}} + c = x^{-2} + \dfrac{3x^4}{4} + \dfrac{3x^{\frac{4}{3}}}{4} + c$

Method notes

Integrate term by term.

You need to remember the constant of integration, c.

Simplifying expressions before integrating

If the expression to be integrated has a product of terms then you will need to use the index laws to simplify the expression before proceeding with the integration process.

Example

Evaluate the following integrals.

a) $\int (2x - 5) x^2 dx$

b) $\int (3x + 2)(x + 1) dx$

c) $\int \dfrac{2x^3 - 5}{x^2} dx$

d) $\int \dfrac{x + 3}{\sqrt{x}} dx$

Answer

a) $\int (2x - 5) x^2 dx = \int 2x^3 - 5x^2 dx$

$$= 2\dfrac{x^4}{4} - 5\dfrac{x^3}{3} + c = \dfrac{x^4}{2} - \dfrac{5x^3}{3} + c$$

b) $\int (3x + 2)(x + 1) dx = \int (3x^2 + 5x + 2) dx$

$$= x^3 + 5\dfrac{x^2}{2} + 2x + c$$

c) $\int \dfrac{2x^3 - 5}{x^2} dx = \int \left(\dfrac{2x^3}{x^2} - \dfrac{5}{x^2} \right) dx$

$$= \int (2x - 5x^{-2}) dx = x^2 - 5\dfrac{x^{-1}}{-1} + c = x^2 + 5x^{-1} + c$$

d) $\int \dfrac{x + 3}{\sqrt{x}} dx = \int \dfrac{x}{\sqrt{x}} + \dfrac{3}{\sqrt{x}} dx$

$$= \int x^{\frac{1}{2}} + 3x^{-\frac{1}{2}} dx = \dfrac{x^{\frac{3}{2}}}{\frac{3}{2}} + 3\dfrac{x^{\frac{1}{2}}}{\frac{1}{2}} + c = \dfrac{2x^{\frac{3}{2}}}{3} + 6x^{\frac{1}{2}} + c$$

Method notes

In a) and b) multiply out before integrating.

In c) divide out before integrating.

In d) simplify algebraically before integrating.

Method notes

Most questions use x and y but other letters may be used. The same rules apply. Here x and t are the variables.

$$\frac{dx}{dt} = kt^n \rightarrow y = \frac{k}{n+1}t^{n+1} + c$$

$n \neq -1$ and k is a constant.

Using different letters to represent variables

Example

Given that $\dfrac{dx}{dt} = t^2 + 3t + 5$ find an expression for $x(t)$.

Answer

$$\frac{dx}{dt} = t^2 + 3t + 5 \implies x = \frac{t^3}{3} + 3\frac{t^2}{2} + 5t + c = \frac{t^3}{3} + \frac{3t^2}{2} + 5t + c$$

Finding the equation of a curve given the gradient function

When evaluating an indefinite integral you must include a constant of integration. For example,

$\dfrac{dy}{dx} = x \rightarrow y = \dfrac{x^2}{2} + c$. For each value of c there is a different graph of y against x. In principle there are in fact an infinite number of parallel curves giving a 'family' of curves as shown by the graphs in Figure 5.2.

Fig. 5.2
The graphs of the family of curves.

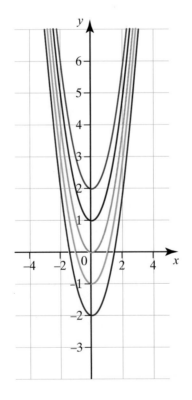

If we are told that the curve passes through a particular point, then we are able to find the value of the constant of integration which will enable us to find the particular curve or 'family member' given in the question.

Example

Find the equation of the curve with gradient function $\dfrac{dy}{dx} = 3x^2 + \dfrac{2}{x^3} - 5$

that passes through the point (2, 1).

Answer

$\dfrac{dy}{dx} = 3x^2 + \dfrac{2}{x^3} - 5 = 3x^2 + 2x^{-3} - 5$

$y = x^3 + 2\dfrac{x^{-2}}{-2} - 5x + c = x^3 - x^{-2} - 5x + c$

The graph of this function passes through the point (2, 1).

Substitute $y = 1$ and $x = 2$ into the expression for y.

$1 = 2^3 - 2^{-2} - 5 \times 2 + c = 8 - \dfrac{1}{4} - 10 + c$

$c = 3\dfrac{1}{4}$

So the equation of the curve (the particular 'family member') is

$y = x^3 - x^{-2} - 5x + 3\dfrac{1}{4}$

Method notes

Integrate in the normal way remembering the constant of integration.

Use the given information that $y = 1$ when $x = 2$ to solve for c.

Example

Given that $\dfrac{dx}{dt} = 3t^2 + 2t + 5$ and that $x = 3$ when $t = 1$ find the value of x when $t = 2$.

Answer

$\dfrac{dx}{dt} = 3t^2 + 2t + 5$

$x = t^3 + t^2 + 5t + c$

$x = 3$ when $t = 1 \Rightarrow 3 = 1^3 + 1^2 + (5 \times 1) + c = 7 + c$

$\Rightarrow c = -4$

$x = t^3 + t^2 + 5t - 4$

When $t = 2$, $x = 2^3 + 2^2 + (5 \times 2) - 4 = 18$

Method notes

x and t are the variables in this question.

Integrate for $x(t)$ in the normal way remembering the constant of integration.

Use the given information $x = 3$ when $t = 1$ to find the constant c.

Write out the formula for $x(t)$.

Substitute for $t = 2$ to find the corresponding value of x.

Stop and think answer

$\int c\,dx$ rewritten means $\int cx^0 dx$

but since $x^0 = 1$ Using rules of indices

then $\int cx^0 dx = \dfrac{cx^{0+1}}{0+1} + m$ Using the rule of integration

where m is the constant of integration

so $\int c\,dx = cx + m$

Questions

You cannot use a calculator.
A formula sheet is attached for your reference.

1. (a) Write down the value of $8^{\frac{1}{3}}$ (1)

   (b) Find the value of $8^{-\frac{2}{3}}$ (2)

   (c) Simplify $(8y^{15})^{\frac{2}{3}}$ (2)

2. (a) Simplify $\dfrac{9 + \sqrt{5}}{1 - \sqrt{5}}$ giving your answer in the form $x + y\sqrt{5}$ (3)

   (b) Give the values of x and y as fraction forms. (1)

3. (a) Find the set of values of x for which $7 < 2x + 7 < 17$
      where x is an integer. (3)

   (b) (i) Find the values of q and r if $2x^2 + 12x + 7 = 2(x + q)^2 + r$

      (ii) Find the set of values of x for which $2x^2 + 12x + 7 < 0$ (5)

4. Solve the simultaneous equations

   $x + y = 1$

   $x^2 + 3y = 7$ (6)

5. (a) Sketch the graph of the curve with equation $y = x^2 - 2x - 3$
      State the coordinates of the points of intersection of the
      curve with the axes. (3)

   (b) Write down the equation of the axis of symmetry for this graph. (1)

   (c) Using algebra find the x-coordinates of the points of intersection
      of the curve with equation $y = x^2 - 2x + 3$ and the line
      $y = x + 2$ giving your answer in exact form. (5)

6. (a) State the next three terms in the sequence $-1, 4, 9, \ldots$ (2)

   (b) Give a formula for the general term u_n of this sequence. (1)

   (c) Find the sum of the first 20 terms in this sequence. (3)

7. Figure 1 shows a sketch of the curve with equation $y = f(x)$.
   The curve crosses the x-axis at the points $(1, 0)$ and $(2, 0)$ and the
   turning point is $(1.5, 1.25)$. The graph crosses the y-axis at $(0, -10)$.

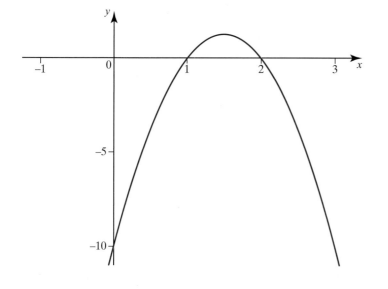

Fig. 1
Graph of $y = f(x)$.

On separate diagrams **sketch** the following functions

(a) $f(-x)$ (3)

(b) $3f(x)$ (2)

(c) $f(x-1)$ (3)

On each diagram show clearly the co-ordinates of any intercept points with the x-axis and the coordinates of the turning point.

8. A line L passes through the two points $(-2, -3)$ and $(2, 5)$.

(a) Find the equation of L in the form $ax + by + c = 0$

State the values of a, b and c.   (6)

(b) Find the equation of the line which is perpendicular to L and which also passes through the point $(2, 5)$.   (4)

9. Given that $y = f(t) = 2t^3 - t - \dfrac{4}{t^2}$, $t \neq 0$

(a) find $f'(t)$ (3)

(b) find $f''(t)$ (1)

(c) find $\int f(t)\,dt$ (4)

10. The point P $(1, -5)$ lies on the curve C with equation $y = x^2 + 2x - 8$

(a) Find the equation of the tangent to the curve at P giving your answer in the form $y = mx + c$ where m and c are constants.   (7)

(b) Find the equation of the normal to the curve at P giving your answer in the form $y = mx + c$ where m and c are constants.   (4)

Answers

1. (a) $8^{\frac{1}{3}}$ means find the cube root of 8 (using the rules of indices) so the answer is 2   (1)

   (b) $8^{-\frac{2}{3}}$ rewrite this as $(8^{\frac{1}{3}})^{-2}$ using rules of indices   (2)

   $$\Rightarrow (2)^{-2} \text{ from (a)}$$
   $$\Rightarrow \frac{1}{2^2} \text{ using rules of indices}$$
   $$= \frac{1}{4}$$

   (c) $(8y^{15})^{\frac{2}{3}}$ rewrite this as $(8^{\frac{2}{3}}y^{\frac{15\times 2}{3}})$ using rules of indices   (2)

   $$\Rightarrow (2^2 y^{10}) \text{ using rules of indices}$$
   $$= 4y^{10}$$

2. (a) Rationalising the denominator gives   (3)

   $$\frac{9 + \sqrt{5}}{1 - \sqrt{5}} = \frac{(9 + \sqrt{5})(1 + \sqrt{5})}{(1 - \sqrt{5})(1 + \sqrt{5})} = \frac{9 + \sqrt{5} + 9\sqrt{5} + 5}{1 - 5}$$
   $$= -\frac{14}{4} - \frac{10\sqrt{5}}{4} = -\frac{7}{2} - \frac{5\sqrt{5}}{2}$$

   (b) $x = -\dfrac{7}{2} \quad y = -\dfrac{5}{2}$ (1)

3. (a) Take the question in two sections: $7 < 2x + 7$ and $2x + 7 < 17$ (3)

   For $7 < 2x + 7$ subtract 7 from both sides $\Rightarrow 0 < 2x$ so $x > 0$

   For $2x + 7 < 17$ subtract 7 from both sides $\Rightarrow 2x < 10$ so $x < 5$

   Combining the two answers gives $0 < x < 5$ and since x is an integer then $x = 1, 2, 3, 4$

   (b) (i) $2x^2 + 12x + 7 = 2(x^2 + 6x) + 7$ (5)

   Using the method of completing the square
   $2(x^2 + 6x) + 7 = 2(x + 3)^2 - 18 + 7$

   Simplifying gives $2x^2 + 12x + 7 = 2(x + 3)^2 - 11 \Rightarrow q = 3$
   and $r = -11$

   (ii) From (b) $2x^2 + 12x + 7 < 0 \Rightarrow 2(x + 3)^2 - 11 < 0$

   $\Rightarrow 2(x + 3)^2 < 11 \Rightarrow (x + 3)^2 < 5.5$

   Taking the square root gives two possible boundary values
   $+\sqrt{5.5}$ and $-\sqrt{5.5}$ for $(x + 3)$

   $\Rightarrow -\sqrt{5.5} < (x + 3) < \sqrt{5.5}$

   giving the answer $-\sqrt{5.5} - 3 < x < \sqrt{5.5} - 3$

4. Label the equations $x + y = 1$ (1)

$$x^2 + 3y = 7 \qquad (2)$$

Rewriting (1) gives $y = 1 - x$ (3)

Substitute (3) in (2) $\Rightarrow x^2 + 3(1 - x) = 7$

simplifying $\Rightarrow x^2 - 3x - 4 = 0$

Factorising $\Rightarrow (x - 4)(x + 1) = 0$

$\Rightarrow x = 4$ or $x = -1$

When $x = 4$ in (1) $y = -3$ and when $x = -1$ in (1) $y = 2$

giving the solutions $(4, -3)$ and $(-1, 2)$.

5. (a)

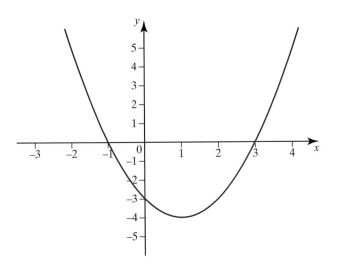

Fig. 2
Graph of $y = x^2 - 2x - 3$

Points of intersection with the x-axis are $(-1, 0)$, $(3, 0)$ and with the y-axis is $(0, -3)$.   (3)

(b) The axis of symmetry is the line $x = 1$ (half way between $x = -1$ and $x = 3$, which are the x values where the curve crosses the x-axis).   (1)

(c) Points of intersection are where the curve and the line cross.   (5)

This will be where the y-coordinate on the curve $y = x^2 - 2x - 3$ (1)

and the y-coordinate on the line $y = x + 2$ (2) are both equal

Solving (1) and (2) as simultaneous equations $\Rightarrow$ Substitute (2) in (1)

$\Rightarrow x^2 - 2x - 3 = x + 2$

$x^2 - 3x - 5 = 0$

Using the quadratic formula with $a = 1, b = -3, c = -5$

$$\Rightarrow x = \frac{3 \pm \sqrt{9 + 20}}{2} = \frac{3 \pm \sqrt{29}}{2}$$

6. (a) Given the sequence $-1, 4, 9$ .... (2)

We can see that it is an arithmetic sequence with first term $a = -1$ and common difference $d = 5$

$\Rightarrow$ the next term will be $9 + 5 = 14$, then $14 + 5 = 19$ and the next $19 + 5 = 24$

$\Rightarrow$ the next three terms will be 14, 19, 24

(b) The general or nth term of an arithmetic sequence is given by
$$u_n = a + (n-1)d \qquad (1)$$

as $a = -1$ and $d = 5$ in this sequence $\Rightarrow$
$$u_n = -1 + (n-1)5 \Rightarrow u_n = -1 + 5n - 5$$
$$\Rightarrow u_n = 5n - 6$$

(c) The sequence is an arithmetic sequence with $a = -1$ and $d = 5$ (3)

General formula for sum of n terms is: $S_n = \dfrac{n}{2}(2a + (n-1)d)$

We require the sum of the first 20 terms so $n = 20$

Therefore sum of first 20 terms $= \dfrac{20}{2}(2a + (n-1)d)$
$$= 10(-2 + 19 \times 5) = 930$$

7. (a) $y = f(-x)$ will be a reflection in the y-axis of $y = f(x)$ as shown in the Figure 3. (3)

(b) $y = 3f(x)$ will be a stretch, scale factor 3 in the y direction of $y = f(x)$ as shown in Figure 4. (2)

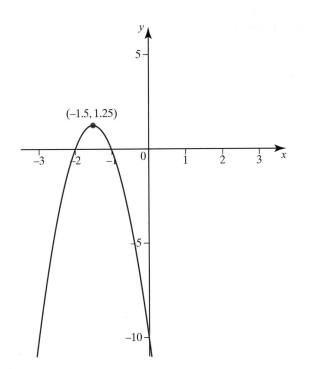

Fig. 3
Graph of $y = f(-x)$

Exam tips

To gain full marks for parts (a), (b) and (c) both coordinates of the turning points must be stated and the points of intersection with the axes must also be correct.

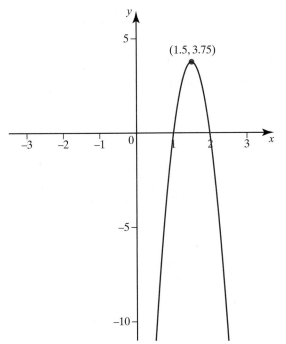

Fig. 4
Graph of $y = 3f(x)$

(c) $y = f(x - 1)$ will be a translation of $f(x)$, 1 unit to the right parallel to the x-axis as shown in Figure 5 below. (3)

Fig. 5
Graph of $y = f(x - 1)$.

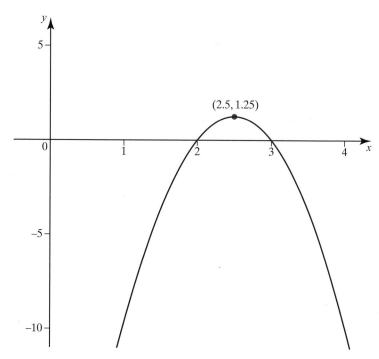

8. (a) Using the formula $\dfrac{y_2 - y_1}{x_2 - x_1}$ to find the gradient between 2 points

   $(-2, -3)$ and $(2, 5)$ gradient of $L = \dfrac{5 - (-3)}{2 - (-2)} = \dfrac{8}{4} = 2$ (6)

   Using the formula of a straight line given a point on it and its

   gradient $\dfrac{y - y_1}{x - x_1} = m$

   $\Rightarrow$ equation of the line L is $\dfrac{y - 5}{x - 2} = 2 \Rightarrow y - 5 = 2(x - 2)$

   Rewriting $\Rightarrow y - 5 = 2x - 4 \Rightarrow 2x - y + 1 = 0$ which is the equation of line L.

   Comparing this with the general equation of a line $ax + by + c = 0$
   $\Rightarrow a = 2, b = -1, c = 1$

(b) Using the fact that the gradients of perpendicular lines multiply together to give -1 and the gradient of $L = 2$ then the gradient of a line perpendicular to L is $-\dfrac{1}{2}$ (4)

Using the general equation $\dfrac{y - y_1}{x - x_1} = m$

$\Rightarrow$ equation of the perpendicular to L through $(2, 5)$ is $\dfrac{y - 5}{x - 2} = -\dfrac{1}{2}$

rewriting $\Rightarrow 2(y - 5) = -(x - 2)$

$\Rightarrow 2y - 10 = -x + 2$

$\Rightarrow x + 2y - 12 = 0$ which is the equation of the line perpendicular to L which passes through the point $(2, 5)$.

9. Given $f(t) = 2t^3 - t - \dfrac{4}{t^2}$ to find $f'(t)$ means we have to differentiate with respect to t.

(a) Rewriting using rules of indices $f(t) = 2t^3 - t - \dfrac{4}{t^2} \Rightarrow f(t) = 2t^3 - t - 4t^{-2}$
using the rule of differentiation (3)

$f'(t) = 6t^2 - 1 - 4 \times (-2t^{-3})$

Rewriting using rules of indices $\Rightarrow f'(t) = 6t^2 - 1 + \dfrac{8}{t^3}$

(b) To find $f''(t)$ means we have to differentiate $f'(t)$ with respect to t. (1)

$f'(t) = 6t^2 - 1 + \dfrac{8}{t^3}$ which rewritten $\Rightarrow f'(t) = 6t^2 - 1 + 8t^{-3}$

Using the rule of differentiation $\Rightarrow f''(t) = 12t + 8(-3t^{-4})$

$f''(t) = 12t - \dfrac{24}{t^4}$

(c) Given $f(t)$ to find $\int f(t)dt$ means we have to integrate $f(t)$ with respect to t. (4)

Rewriting $f(t)$ using the rules of indices

$\Rightarrow \int f(t)dt = \int 2t^3 - t - 4t^{-2}dt$

Using the rule of integration

$\int 2t^3 - t - 4t^{-2}dt = 2\dfrac{t^4}{4} - \dfrac{t^2}{2} - 4\dfrac{t^{-1}}{-1} + c$ where c is the constant of integration

Rewriting using rules of indices $\Rightarrow \int f(t)\,dt = \dfrac{t^4}{2} - \dfrac{t^2}{2} + \dfrac{4}{t} + c$

10. (a) $y = x^2 + 2x - 8$

    Differentiating y with respect to x will give the gradient function i.e. the gradient of the tangent to the curve at any point on the curve. (7)

    Using the rule of differentiation if $y = x^2 + 2x - 8 \Rightarrow \dfrac{dy}{dx} = 2x + 2$

    therefore at the point $(1, -5)$ $x = 1$ and $y = -5 \Rightarrow \dfrac{dy}{dx} = 2(1) + 2 = 4$

    So the gradient of the tangent to the curve at the point $(1, -5) = 4$

    Using the general formula $\dfrac{y - y_1}{x - x_1} = m$ for the equation of a line

    the equation of the tangent to the curve at the point $(1, -5)$ is

    $\dfrac{y - (-5)}{x - 1} = 4$

    $\Rightarrow \dfrac{y + 5}{x - 1} = 4 \Rightarrow y + 5 = 4(x - 1)$

    Simplifying gives $y + 5 = 4x - 4$

    $\Rightarrow y = 4x - 9$ which is the equation of the tangent to the curve.

    (b) The normal at a point on a curve is a line which is perpendicular to the tangent to the curve at that point. We also know that the gradients of perpendicular lines multiply together to give -1 (4)

    $\Rightarrow$ gradient of tangent $\times$ gradient of normal $= -1$

    $4 \times$ gradient of normal $= -1 \Rightarrow$ gradient of normal $= -\dfrac{1}{4}$

    Using the general formula $\dfrac{y - y_1}{x - x_1} = m$

    equation of the normal to the curve at the point $(1, -5)$ is

    $\dfrac{y - (-5)}{x - 1} = -\dfrac{1}{4}$

    $\Rightarrow \dfrac{y + 5}{x - 1} = -\dfrac{1}{4} \Rightarrow 4y + 20 = -1(x - 1)$

    $\Rightarrow 4y + 20 = -x + 1 \Rightarrow y = -\dfrac{1}{4}x - \dfrac{19}{4} = 0$ which is the

    equation of the normal.

Set notation

$\in$	is an element of
$\notin$	is not an element of
$\{x_1, x_2, \dots\}$	the set with elements $x_1, x_2, \dots$
$\{x : \dots\}$	the set of all x such that $\dots$
$n(A)$	the number of elements in set A
$\varnothing$	the empty set
ε	the universal set
A'	the complement of the set A
$\mathbb{N}$	the set of natural numbers, $\{1, 2, 3, \dots\}$
$\mathbb{Z}$	the set of integers, $\{0, \pm 1, \pm 2, \pm 3, \dots\}$
$\mathbb{Z}^+$	the set of positive integers, $\{1, 2, 3, \dots\}$
$\mathbb{Z}_n$	the set of integers modulo n, $\{0, 1, 2, \dots, n-1\}$
$\mathbb{Q}$	the set of rational numbers, $\left\{\dfrac{p}{q} : p \in \mathbb{Z}, q \in \mathbb{Z}^+\right\}$
$\mathbb{Q}^+$	the set of positive rational numbers, $\{x \in \mathbb{Q} : x > 0\}$
$\mathbb{Q}_0^+$	the set of positive rational numbers and zero, $\{x \in \mathbb{Q} : x \geq 0\}$
$\mathbb{R}$	the set of real numbers
$\mathbb{R}^+$	the set of positive real numbers $\{x \in \mathbb{R} : x > 0\}$
$\mathbb{R}_0^+$	the set of positive real numbers and zero, $\{x \in \mathbb{R} : x \geq 0\}$
$\mathbb{C}$	the set of complex numbers
(x, y)	the ordered pair x, y
$A \times B$	the cartesian product of sets A and B, ie $A \times B = \{(a, b) : a \in A, b \in B\}$
$\subseteq$	is a subset of
$\subset$	is a proper subset of
$\cup$	union
$\cap$	intersection
$[a, b]$	the closed interval, $\{x \in \mathbb{R} : a \leq x \leq b\}$
$[a, b), [a, b[$	the interval $\{x \in \mathbb{R} : a \leq x < b\}$
$(a, b],]a, b]$	the interval $\{x \in \mathbb{R} : a < x \leq b\}$
$(a, b),]a, b[$	the open interval $\{x \in \mathbb{R} : a < x < b\}$
$y R x$	y is related to x by the relation R
$y \sim x$	y is equivalent to x, in the context of some equivalence relation

Miscellaneous symbols

$=$	is equal to
$\neq$	is not equal to
$\equiv$	is identical to or is congruent to
$\approx$	is approximately equal to
$\cong$	is isomorphic to
$\propto$	is proportional to
$<$	is less than
$\leq$	is less than or equal to, is not greater than
$>$	is greater than
$\geq$	is greater than or equal to, is not less than
∞	infinity
$p \wedge q$	p and q
$p \vee q$	p or q (or both)
$\sim p$	not p
$p \Rightarrow q$	p implies q (if p then q)
$p \Leftarrow q$	p is implied by q (if q then p)
$p \Leftrightarrow q$	p implies and is implied by q (p is equivalent to q)
$\exists$	there exists
$\forall$	for all

Operations

$a + b$	a plus b
$a - b$	a minus b
$a \times b,\ ab,\ a.b$	a multiplied by b
$a \div b,\ \dfrac{a}{b},\ a/b$	a divided by b
$\displaystyle\sum_{i=1}^{n} a_i$	$a_1 + a_2 + \ldots + a_n$
$\displaystyle\prod_{i=1}^{n} a_i$	$a_1 \times a_2 \times \ldots \times a_n$
$\sqrt{a}$	the positive square root of a
$|a|$	the modulus of a
$n!$	n factorial
$\dbinom{n}{r}$	the binomial coefficient $\dfrac{n!}{r!(n-r)!}$ for $n \in \mathbb{Z}^+$ $\dfrac{n(n-1)\ldots(n-r+1)}{r!}$ for $n \in \mathbb{Q}$

Functions

f(x)	the value of the function f at x
f : A $\rightarrow$ B	f is a function under which each element of set A has an image in set B
f : $x \rightarrow y$	the function f maps the element x to the element y
f^{-1}	the inverse function of the function f
g$\circ$f, gf	the composite function of f and g which is defined by $(g \circ f)(x)$ or $gf(x) = g(f(x))$
$\lim_{x \to a}$ f(x)	the limit of f(x) as x tends to a
Δx, dx	an increment of x
$\dfrac{dy}{dx}$	the derivative of y with respect to x
$\dfrac{d^n y}{dx^n}$	the nth derivative of y with respect to x
f$'(x)$, f$''(x)$,..., f$^{(n)}(x)$	the first, second, ..., nth derivatives of f(x) with respect to x the indefinite integral of y with respect to x
$\int y\,dx$	the indefinite integral of y with respect to x
$\int_a^b y\,dx$	the definite integral of y with respect to x between the limits $x = a$ and $x = b$
$\dot{x}, \ddot{x}, \ldots$	the first, second, ... derivatives of x with respect to t

Exponential and logarithmic functions

$\log_a x$	logarithm to the base a of x
$\lg x$, $\log_{10} x$	logarithm of x to base 10

Circular and hyperbolic functions

sin, cos, tan, cosec, sec, cot	the circular functions

Formulae you need to remember

Quadratic equations

The roots of $ax^2 + bx + c = 0$ are $x = \dfrac{-b \pm \sqrt{b^2 - 4ac}}{2a}$

Differentiation
The derivative of x^n is nx^{n-1}

Integration

The integral of x^n is $\dfrac{1}{n+1}x^{n+1} + c, n \neq -1$

Formulae given in the formulae booklet

Arithmetic sequences
Formula for the nth term $u_n = a + (n-1)d$ where a is the first term and d is the common difference

Arithmetic series
Sum of the first n terms
$$S_n = \tfrac{1}{2}n(a + l) = \tfrac{1}{2}n[2a + (n-1)d]$$

Mensuration
Surface area of a sphere $= 4\pi r^2$

Area of the curved surface of a cone $= \pi r \times$ slant height